Behavioral Ecology

Kleber Del Claro

Behavioral Ecology

Initial Dialogues and Approaches

Springer

Kleber Del Claro
Instituto de Biologia
Universidade Federal de Uberlandia
Uberlândia, Brazil

ISBN 978-3-032-13987-0 ISBN 978-3-032-13988-7 (eBook)
https://doi.org/10.1007/978-3-032-13988-7

The original submitted manuscript has been translated into English. The translation was done using artificial intelligence. A subsequent revision was performed by the author(s) to further refine the work and to ensure that the translation is appropriate concerning content and scientific correctness. It may, however, read stylistically different from a conventional translation.

This Springer imprint is published by the registered company Springer Nature Switzerland AG
The registered company address is: Gewerbestrasse 11, 6330 Cham, Switzerland

Preface

Hello, colleague! Welcome to the pages of my new book, where we will discuss a beautiful, captivating, innovative, and evolutionarily fascinating science, full of inspiring stories—*Behavioural Ecology*—and the use of *Animal Behaviour* tools, which, in studies of biotic interactions, often reveal the true naturalist within each of us who loves this field of science.

This "written conversation" is by no means intended to be a comprehensive book on Behavioural Ecology, covering all its various aspects, approaches, statistical perspectives, and so on. This book is aimed at those who teach Animal Behaviour and Behavioural Ecology. It is directed at undergraduate and postgraduate students who may be interested in this area of science, which has significantly driven human development and made us who we are.

Teaching and lecturing on animal behaviour and its most fascinating branch, Behavioural Ecology, have been a constant activity throughout my academic life. Since 1992, I have taught the foundations of the field and how to work with ecology and behaviour at various public institutions in Brazil and abroad. Over the years, introducing young students to this area of science, I have had the pleasure of meeting wonderful people, many of whom are now active teachers and researchers in this field. These people have allowed me to conclude that, with dedication and enthusiasm, this area offers an excellent scientific path, good prospects for a permanent career, and great opportunities for internationalization. I strongly encourage biologists, veterinarians, animal scientists, and experimental psychologists to pursue this path. The chances of professional success and fulfilment are very good in this remarkable field, which is constantly evolving and innovating.

Our conversation will cover the history and importance of behavioural studies. We will discuss the main tools, including a special one—human curiosity—and your own ability, reader, to identify a potential study and know how to develop it. We will not neglect the basic techniques or your theoretical preparation, nor your development in learning to work with behaviour and behavioural ecology. We will talk about methods, limitations and possibilities, planning and hypotheses, always sharing stories, with examples from our own experience and those close to you. Yes, the intention is for this to be an engaging and accessible read, one that sparks curiosity, presents opportunities, opens doors, and does not stifle scientific ambition; that allows you to dream. After all, life punishes those who have wings but do not fly; and as I believe you may have strong wings, my aim is to encourage your flight, pushing you towards a sky full of possibilities. That is how life should be, and science as a part of it.

Welcome to our home! The door is open!

Uberlândia, Brazil

Kleber Del Claro

Acknowledgements

I sincerely thank all those who have dedicated their time to teaching me. My heartfelt thanks to each of my mentors. All of you are of incomparable importance.

I am grateful to each student I have mentored, from whom I have learned and by whom I have been inspired, as well as for the pleasure and happiness of seeing them grow, progress, and find their own path.

I am grateful to the Federal University of Uberlândia (UFU), where I have worked since 1992, and to all its staff and academics. I thank the Publisher of the Federal University of Uberlândia (Edufu) and its management team, the editorial board, and the general editor for their support and trust. This book was first published in Portuguese by EDUFU/ED. UFMG—*Ecologia Comportamental: uma conversa sobre biologia evolutiva*—ISBN 978-65-88055-15-1, in 2024, who kindly permitted its translation into English. I would like to thank the entire editorial board of EDUFU on behalf of Eduardo Moraes Warpechowski and Sertório de Amorim e Silva Neto. I would also like to thank Springer/Nature, and especially Mrs. Gabriela Capille, Editor of Life Sciences, Latin America and the Caribbean, for her support and enthusiasm in believing in my work.

I thank the institutions that have supported me at various points in my career, or throughout it; the Federal Universities of Uberlândia, Juiz de Fora, Mato Grosso do Sul, Paraná, and the Faculty of Philosophy, Sciences and Letters of Ribeirão Preto (FFCLRP-USP). I am grateful for all the support I have received from the Conselho Nacional de Desenvolvimento Científico e Tecnológico (CNPq) since 1996, as a PQ 1A researcher. To the Coordenação de Aperfeiçoamento de Pessoal de Nível Superior (Capes) and again to CNPq, for the scholarships and project funding granted to me, my collaborators, and my students.

I would like to thank Denise Lange, Everton Tizo-Pedroso, José Henrique Pezzonia, Raoni Rosa Rodrigues, and Helena Maura Torezan-Silingardi, who kindly each provided a photo that I used in the book.

I thank my three sons, Vergílio, Augusto, and Túlio, for always being a source of joy and pride, and for always supporting and motivating me. Raising three honourable, upright men of character in any era has always been a challenging task. My loves, always.

I thank my partner since 1983, Maura, for always supporting me, for offering criticism when needed, and for encouraging me at every moment. Always the love of my life.

I thank nature, for the opportunity to live alongside its creatures, interactions, and images.

Finally, I thank you, for dedicating a precious part of your life, of your time, to read the words that make up this conversation. May it help you to take flight!

This book is dedicated to teachers who genuinely care about their students in all aspects of their lives. As an example, Prof. Dr Helena Maura Torezan Silingardi.

Contents

Chapter 1
It All Began Out of Necessity and Curiosity

Observation, inquisitive and critical analysis, and the study of the behaviour of other animals have been fundamental to our success as a species.

Since the dawn of humanity, the observation and understanding of animal behaviour have played a vital role in the development of the human species. This fascinating journey of discovery has provided us with valuable *insights* that have driven our survival, evolution, and progress over the centuries. Imagine yourself in the distant past, when our ancestors faced the challenges of the wild. It was through observing animal behaviour that they uncovered the best hunting strategies, learned about their environment, found sources of food, and protected themselves from relentless predators. By watching the animals around them, our ancestors acquired essential knowledge for their own survival.

But it did not stop there. As time went on, the study of animal behaviour brought us significant advances in various fields. Agriculture and animal domestication became possible thanks to careful observation of plant growth patterns and an understanding of animal behaviour. Our early explorers learned which seeds were most promising and which animals could be domesticated to assist with agricultural tasks, transport, and even to meet their nutritional needs. Furthermore, the study of animal behaviour has been a true boon for medicine and scientific research. Through meticulous observation of animals, scientists have deciphered the secrets of physiology, genetics, and the functioning of biological systems. This in-depth understanding has led to innovative medical treatments and remarkable scientific advances, such as those achieved in neuroscience and psychology. And we must not forget the exciting fact that animal behaviour can reveal much about ourselves as human beings! By studying animals, we can find parallels and influences in our own behaviour. This approach allows us to better understand human evolution, comprehend our basic instincts, and even develop treatments for psychological disorders.

Thus, the journey of *Homo sapiens* through the study of animal behaviour is an exciting adventure, full of discoveries that have shaped the course of human history. These valuable *insights* have enabled our survival in challenging environments,

K. Del Claro, *Behavioral Ecology*, https://doi.org/10.1007/978-3-032-13988-7_1

unlocked the secrets of agriculture and domestication, driven medical and scientific advances, and provided profound insight into ourselves as human beings. Therefore, I invite you to embark on this thrilling journey of discovery and understanding of the animal kingdom, because its importance to human development is truly extraordinary!

1.1 Fragile Creatures with Remarkable Cognitive Development

Humans are fragile creatures and can be killed or fatally injured by a range of animals that still exist today. They may attack us intentionally in territorial defense, in self-defense, for food, or accidentally—out of panic, or simply because we happened to encounter them. Yes, imagine an elephant with its calf being startled by you on a path, or the same happening with a polar bear or a jaguar. Or a great white shark, which does not know you are not a seal. Picture yourself fishing unwittingly in a river with crocodiles, splashing your feet in the water. Imagine stepping on a rattlesnake or shaking a vine that supports a huge hive of Africanized bees. In short, there are a thousand and one ways for a human to be killed, intentionally or accidentally, in nature. And there is little one can do in "hand-to-hand" defense, as we lack strong claws, powerful teeth, or a punch capable of felling a bear. Not even running is enough for humans to escape from large felines, as a leopard can cover 100 m in 10 s with ease (Box 1.1).

Box 1.1 Accidentally or intentionally, since we exist, we have been vulnerable

Human deaths by animals in 2022

Snakes	100,00 deaths
Domestic dogs	30,000 deaths
Crocodiles	1000 deaths
Hippos	500 deaths
Elephants	500 deaths
Lion	250 deaths
African buffaloes	200 deaths

Excluding disease transmitters such as mosquitoes, snails, and worms.

https://www.statisa.com/statistics/448169/deadliest-creatures-in-th-world-by-number-of-human-deaths/

The human species was relatively easy prey for numerous species of felines and canids that hunted in groups or alone. But it was also prey for other primates, including its own species.

Moreover, when we first descended from the trees to the ground and began social life, we knew little about which plants could be eaten, barely mastered any kind of tool, and our communication was still rudimentary. Yes, there is no doubt, there was a time when we lived on berries, small prey, eggs, and leftovers left by other animals. But then came the various cognitive revolutions that gradually enabled our ancestors to observe the lives of the animals around them with greater attention. What did they eat? How did they hunt and how did they hunt us? Where and when did they sleep? How did they defend themselves? Did they live alone or in groups?

Cognitive development, the awakening of human curiosity, the ability to observe, analyse, experiment, and, later on, to synthesise and transmit learning to others—the basic components of science—were born from the human need to observe animal behaviour for survival. Our interest in animals is rooted in our origins. This is easily demonstrated through art: cave paintings and engravings existed as early as 40,000 BC, etched onto the walls or ceilings of shelters or caves. In Africa, Asia, Oceania, Europe, and the Americas, including Brazil, there are countless records of animal observation by prehistoric humans. In the Spanish cave of Altamira, also known as the "Sistine Chapel of Prehistory," the cave paintings are striking for their realism and the representation of animals, humans, and their interactions. There are numerous paintings of animals, including some that have been shot with arrows or cornered by humans. The same is true in many other caves scattered across the world. These engravings they provide physical evidence of the earliest intergenerational transmissions of the recording and study of human observation of animal behaviour. In Brazil, we can observe several of these remarkable examples of prehistoric art, for instance, in the Serra da Capivara National Park, in Piauí, our main open-air collection, declared a World Cultural Heritage Site by UNESCO. However, a lesser-known site also houses very ancient paintings. In Goiás, the caves and grottos of the Serra das Araras, near Serranópolis, are striking for their simplicity and informational content. The details reveal the cognitive development of prehistoric humans and their dedication to studying the animals around them. On the walls of this cave in Serranópolis, dating from approximately 11,000 BC to 12,000 BC, prehistoric humans recorded the development of an amphibian, possibly a frog. There are figures resembling tadpoles alongside others in which legs can be seen. Elsewhere in the cave, we find depictions of fish followed by alternating lines that may indicate the presence or absence of these prey at certain times. Thus, for over 12,000 years, humans have been studying the animals of the Cerrado, their behaviour, and their interactions (Fig. 1.1).

The development of cognition, with the capacity for synthesis and prediction—that is, the imagination or visualisation of future possibilities—gave our species a tremendous advantage. The *Homo sapiens* of the post-cognitive revolution began not only to observe but also to hypothesise; they observed other animals and also predicted events based on experimentation, analysis, and synthesis, and were able to transmit this knowledge. First verbally, then through drawings, and finally through writing. Necessity, combined with superior intellectual development, with the study of animal and human behaviour as a model, largely triggered human social development and success.

Oral tradition, that is, the transmission of knowledge from one generation to another through speech, storytelling, legends, and songs, was for a long time the main means of perpetuating knowledge acquired from animal observation within the group. This

Fig. 1.1 Two scientists, Professors Helena Maura Torezan-Silingardi and Kleber Del Claro, observing the behaviour of bees (Centridinae) pollinating flowers of Malpighiaceae (*Byrsonima intermedia*) in the Cerrado. These humans observe, conduct experiments to test their hypotheses, analyse and contextualise the results. Subsequently, they transmit this knowledge in the form of written texts or lectures to their peers. It is likely that an ancestor did the same, recounting stories of their observations around a campfire to their social group

simple mode of knowledge transmission was very useful and effective, especially considering that even today a large proportion of the world's population is illiterate or has limited reading skills. Thus, the storytellers of the earliest human settlements were possibly important naturalists, responsible for passing on valuable information to the younger members and to other populations with whom they interacted.

The earliest writings on something akin to paper—papyri—were made from pieces of crushed and pressed sugar cane in 4000 BC to document state affairs, mostly related to the trade of domesticated animals. In ancient documents from various religions, there are numerous accounts of the behaviour and characteristics of domestic animals. But it was in the Ionian colonies of Asia Minor and in Greece, around 500 BC to 300 BC, that the sciences were born, and at the heart of this first Enlightenment were the natural sciences, with studies of animals and plants: how to hunt, how to know where the fish are and their preferences, how to deal with wolves, dogs, oxen, and how to domesticate them. This knowledge spread first throughout the East, then gradually to the rest of the ancient world, mainly via trade routes and wars of conquest.

From the sixteenth century onwards, with the Age of Discovery and the opening of maritime routes to Africa and Asia, as well as the rediscovery of the New World, humanity was swept up in an explosion of information about animal life and behaviour. From the earliest great voyages, shipboard physicians and artists, as well as missionaries, cartographers, and the first naturalists, attracted the attention of rulers and sponsors of these expeditions around the world. In addition to being more pleasant company for the ship's commanders, these individuals were documentarians and, through their writings and paintings, served as the eyes and ears of Europe in the new worlds. One of the most emblematic and valuable documents from this period dates from 1526 and was written by Gonzalo Fernández de Oviedo y Valdés, entitled *Sumario de la natural historia de las Indias*. Valdés, sent to the New World by King Ferdinand of Spain, produced the first detailed description, rich in examples and specific details, of the flora and fauna of Spanish America, especially the region that is now Mexico. The document contained not only descriptions of the behaviour of animals never before seen, but also of the local human populations: what they ate and wore, which animals and plants they domesticated, how they behaved in different circumstances, and what their social organisation was like. All of this—this basic behavioural science—greatly attracted the attention of the European public, eager to learn about new lands, their people, their customs, new animals, foods, and potential medicines.

Thus, countless other contributions were made over the following two centuries to the study of behaviour, and, even if sporadic and anecdotal, they were of enormous importance and laid the rudimentary foundations for the modern study of animal behaviour. Until, at the end of the eighteenth century and the beginning of the nineteenth century, travelling scientists (naturalists), mainly English, German, Dutch, French, and by then some Americans, began to disseminate the results of their expeditions around the world in a more formal and eclectic manner. Most, although also reporting on animals, had a particular interest in geology, geomorphology, ocean currents, and botany, such as Alexander Von Humboldt, Joseph Dalton Hooker, Charles Lyell, and even Charles Robert Darwin, who was first recognised as a great geologist and later as a naturalist. It was the accounts of individuals such as Henry Walter Bates, Alfred Russel Wallace, Fritz Müller, and Charles Robert Darwin that definitively propelled the study of animal behaviour as we know it today. The Englishman Charles Robert Darwin published a series of articles and books, including his volumes on zoology, an extensive study on the biology and taxonomy of barnacles, and, in 1859, his most famous book, *On the Origin of Species*, in which he established one of the most important pillars in the study of Behavioural Ecology to this day: the concept of evolution through the Theory of Natural Selection. According to the author, the traits most advantageous in terms of survival and reproduction would be positively selected by nature to the detriment of less advantageous ones: "…it is not the strongest that survives, but the one that is best adapted…".

In 1871, Darwin published *The Descent of Man*, which, although not specifically a book about behaviour, established important foundations for the future discussions that would arise in the following years regarding the origin of humans and the similarities between the social behaviours of humans and other primates. In 1873, Darwin ventured further and published *The Expression of the Emotions in Man and Animals*, inaugurating what would later become Comparative Psychology. This book is, without doubt, a landmark in the history of the modern study of animal behaviour.

The study of animal and human behaviour, also known as **Ethology** (from the Greek *ethos*—custom, habit), is a multidisciplinary field of knowledge, as it involves aspects of developmental biology, physiology, genetics, evolution, psychology, as well as zoology and ecology. The term ethology emerged in the eighteenth century in France and was used to refer to the description of lifestyles, often overlapping with our current definition of ecological niche. However, in the sense of specifically referring to the study of animal behaviour, the term ethology was first employed in the twentieth century by the Dutch scientist Niko Tinbergen in 1950. Due to pioneering work in the study of behaviour under natural conditions, with a focus on species-specific behavioural patterns, the German Oskar Heinroth and the American Charles Whitman are also regarded as co-founders of modern ethology.

Thus, the study of animal behaviour had a long descriptive beginning, based on isolated descriptions, without a clear methodology and following the tradition of natural history applied to ethology by nineteenth-century naturalists. Only in the second half of the twentieth century did the study of animal behaviour enter a new, more experimental and empirical phase, seeking to understand the evolutionary causes of these behaviours. From the mid-1950s to the late 1960s, three European researchers stood out in this new approach: the Austrians Konrad Zacharias Lorenz and Karl von Frisch, and the Dutchman Niko Tinbergen.

Konrad Lorenz was a zoologist and ornithologist who sought to understand memory in birds. He developed the concept of *imprinting*. In summary: the maternal memory of birds at birth and in their early life. A duckling would follow the duck, its mother, as she was the closest organism at the moment of its birth, constituting an instinctive behaviour. The other Austrian, Karl von Frisch, studied communication in insects, especially bees, having understood and described the function of the different and complex forms of "dance" performed by honey bees, *Apis mellifera*. Niko Tinbergen, in turn, dedicated himself to the study of memory and instinct in animals, from insects to birds. Together, for their *research focused on understanding the organisation and elicitation of individual and social behaviours*, these scientists were awarded the Nobel Prize in Physiology or Medicine in 1973, inaugurating a new era in ethology, one that would lead us to Behavioural Ecology.

1.2 The Evolutionary Trajectory of Animal Behaviour—Ethology

In a **classical definition, Animal Behaviour,** *stricto sensu*, initially comprised three major areas: **Comparative Psychology, Neurobiology**, and **Ethology**. All were much more focused on understanding the causes and the "how" of behaviours (functionality) than their "whys" (history and evolution).

Comparative Psychology was a more descriptive field that initially gave little importance to the evolutionary causes of animal behaviour. The focus in Comparative Psychology is associative learning, especially in humans. In these studies, it is common to use animals such as dogs, as in Ivan Pavlov's classic experiments on conditioning, or Edward Thorndike's cats in his studies of operant conditioning. But in addition to dogs and cats, pigeons and rats were extensively used by Burrhus Frederic Skinner as experimental models to understand cognitive and perceptual processes in humans (Fig. 1.2). Comparative Psychology, with Thorndike, Skinner, and others, was initially centred in the United States.

Neurobiology is based on a biological foundation for understanding certain behaviours. For a long time, this approach focused on understanding the mechanisms underlying the functioning of the nervous system and its behavioural responses, often to the detriment of evolutionary considerations. Currently, it includes lines of research with a phylogenetic interest, seeking to understand the origins of neural system formation and functioning, as well as their behavioural responses. Santiago Ramón y Cajal (1852–1934), a Spanish physician, described the cellular anatomy of the brain and is regarded by many as the father of modern neuroscience. This field is one of the most promising in the study of human behaviour in the twenty-first century. Today, with advances in computational genetics and microscopy, neurobiology aims to study causes and potential treatments for, for example, degenerative brain diseases, neuronal ageing,

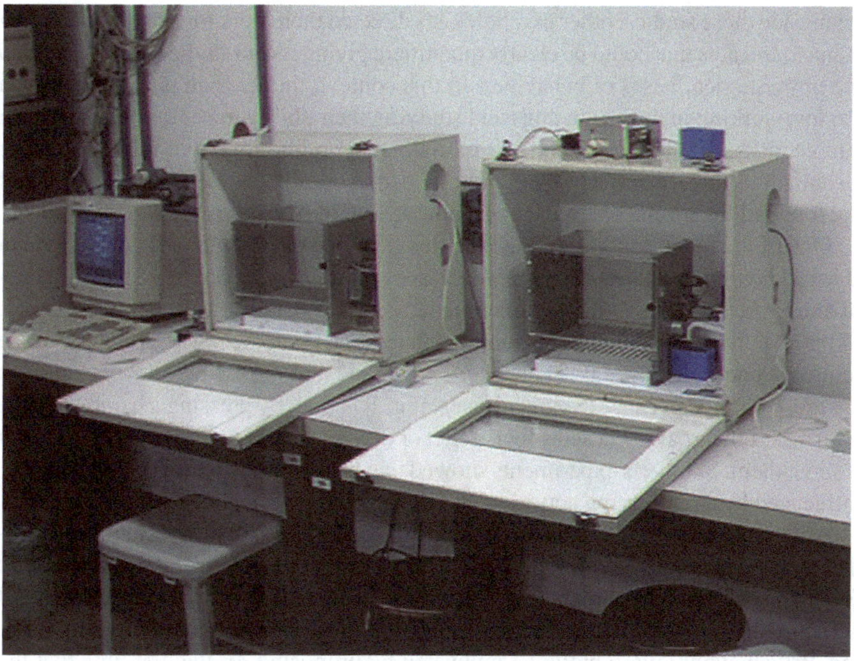

Fig. 1.2 Skinner box—Skinner developed the operant conditioning chamber, which is a variation of Thorndike's puzzle box. Initially, Skinner conducted studies with rats, but later turned to pigeons. The "Skinner box" allows for the observation and manipulation of animal behaviour. In it, the animal is placed inside the chamber and learns to activate levers or respond to light or sound stimuli in exchange for a reward, which may be food or the removal of aversive stimuli, such as a loud alarm. https://free-images.com/display/skinner_boxes_01.html

loss of brain mass, and cognitive decline with ageing. It is an extremely challenging area of immense medical importance. Brazil has internationally renowned figures in this field, such as Suzana Herculano-Houzel and Miguel Nicolelis, among others.

Ethology, the descriptive study of animal behaviour, characterises what we define as "classical ethology," which primarily addressed the physiological bases of behaviour, including causal and functional mechanisms, while placing less emphasis on evolutionary bases (adaptive functions). In the early 1960s, Tinbergen proposed that, first and foremost, behaviour should be observed and described, characterising the *behavioural repertoire* or *ethogram*. After this phase, a comprehensive analysis of behaviour should follow, based on causal, ontogenetic, phylogenetic, and functional analyses.

Given the academic background of the first behaviour researchers, ethology in the twentieth century initially developed within psychology. These early contemporary ethologists were for a long time concerned with stereotyped behaviour patterns, also known as *fixed action patterns* (FAPs). An FAP is any behaviour that can be elicited by a highly characteristic stimulus, known as a sign stimulus or releaser. For example, when a nestling bird opens its beak, displaying the bright colours of its mucosa (yellow, orange, or red) to its mother, it triggers her to regurgitate food for it. Over time, and with the discovery that most behaviours are not, in fact, as stereotyped as initially proposed in modern ethology, FAPs have recently been renamed *modal action patterns* (MAPs). Alongside these studies, other psychologists directed their work towards investigating behavioural acts that could be clearly quantified, giving rise to studies on learning and the physiological bases of behaviour. In this context, the Russian physiologist Ivan Pavlov, working mainly with dogs and studying their ability to be trained to perform simple tasks, developed the concept of classical conditioning, that is, conditioning an animal to perform a specific function. The methodology developed by Pavlov laid the foundations for the emergence of a new school of psychology, *Behaviourism*.

One of the most important figures in this new line of research was the American Burrhus Frederic Skinner, a Harvard professor until 1974, who proposed that the study of animal behaviour should be limited to actions that could be effectively observed. Skinner demonstrated that behavioural patterns that could be rewarded tended to be reinforced and increased in frequency. The classic and most well-known examples are those of mice in cages (Skinner boxes, Fig. 1.2) that, by pressing a bar in the cage with one of their paws, caused a food pellet or drop of water (reward) to be released in front of them. Skinner's experiments showed that behavioural control could be greatly influenced by a reinforcing action.

At the end of the 1960s and the beginning of the 1970s, William D. Hamilton introduced a new approach to ethology, building on the ideas of experimental manipulation (e.g., altering an environmental characteristic to test the function of a behavioural act), initially proposed in the studies of Tinbergen, Von Frisch, and Lorenz. However, this British evolutionary biologist innovated by proposing for the first time that the behaviours exhibited by animals should be studied in terms of their real impact on the adaptive value of species, as reflected in the survival of these animals and their relatives, as well as their ability to leave descendants. In this way, Hamilton proposed the application of natural selection as a tool to understand the genetic foundations shaping behaviours. From his ideas and those of the American entomologist and biologist Edward Osborne Wilson, a new science emerged, **Sociobiology**, which seeks to understand the evolutionary basis for the existence and perpetuation of social behaviours. With Wilson, the concept of fitness, which involves the relative contribution of a phenotype to the next generation compared to other genetic combinations [1], proved to be of vital importance in interpreting animal behaviours.

In the sum of these efforts, what was truly being born was a new way of studying and understanding animal behaviour. This new approach was a revival and refinement of the hypothetico-deductive method (which we will study in detail later), initially employed by the Darwinists, based on the idea of natural selection, reinvigorated by modern knowledge of genetics and evolution. Through experimental manipulation and now also with the use of statistical and mathematical tools, the aim was to understand not only how an animal exhibited a particular behaviour, but, above all, what evolutionary causes maintained this behaviour in terms of alleles present in a population, that is, the adaptive value of behaviours. In this way, Animal Behaviour was transformed into **Behavioural Ecology**.

William Hamilton was able to demonstrate, through his studies with social bees, that the adaptive value of a behaviour can be measured. This is done by calculating the viable offspring (F1) with a chance of future reproductive success that a given behaviour allows the individual to produce over its lifetime. This perspective, combined with that of E. O. Wilson, revolutionised the study of animal behaviour. The redefinition of the concept of adaptation within the context of animal behaviour, together with the understanding of fitness, became a fundamental pillar of Behavioural Ecology, as we can now understand not only how a particular behaviour influenced an individual's survival and reproduction, but also its effects on the populations of these organisms. From then on, it became possible to test and predict the impacts of certain behaviours on producers (plants) and consumers (animals), on biotic interactions, and on the structure of communities and ecological networks. Animal Behaviour, more than just a line of scientific inquiry, became one of the most powerful tools in the multidisciplinary universe of Behavioural Ecology. It contributes decisively to a better understanding of the factors affecting the evolutionary dynamics of natural communities, ecological interaction networks, and biodiversity.

In this new century, an ever-increasing number of scientists are dedicated to the study of Behavioural Ecology. They are spread across the globe; some younger, others older. People just starting their careers and those from the last century are making their final contributions to this grand picture of understanding the animal and human world. It would be absurd to attempt to list here the order of importance of these colleagues' studies and their contributions. In this new and highly dynamic field, the name of the American John Alcock, who passed away in January 2023, deserves special mention. His publication, *Animal Behaviour: An Evolutionary Approach*, now with a new title and a co-author [2], has reached its 11th edition. It is perhaps the contemporary textbook that has most influenced the development of animal behaviour studies among young biologists and ecologists over the past two decades. However, the book that is possibly the most important in the history of Behavioural Ecology had its first edition published in 1978, under the editorship of Professors John R. Krebs and Nicolas B.

Davies. They summarised the advances made in the 1960s and 1970s. It has had several subsequent editions (see [3]).

Between the end of the twentieth century and the beginning of the twenty-first, the advent of the internet and electronic PDFs (*Portable Document Format*) has enabled the increasingly rapid and widespread dissemination of published studies in all fields of knowledge. Not only texts, but also photos and videos of behaviour are now shared by both well-known and anonymous individuals in an attempt to publicise their discoveries. Therefore, be cautious, prudent, and discerning in your internet research, as quantity is not always synonymous with quality. Only through reading and study will you learn to distinguish good from poor scientific communication.

Ernest Mayr defined science as *a small step in the human effort to better understand the world through observation, comparison, experimentation, analysis, synthesis, and contextualisation*. If, in this sense, Biology is a unique science, as it is mutable and adapts to the changes imposed by time, Behavioural Ecology can be considered one of Biology's rebellious offspring. The inquisitive one who attends every party, who unceremoniously dips a finger into the cake's icing. The behavioural ecologist of the twenty-first century is the nineteenth-century naturalist equipped with all the knowledge accumulated by Biology over the past two centuries, from genetics to ecology, including physiology, zoology, and botany. They make use of the most modern electronic equipment for documentation and analysis, including all manner of statistical packages, but never dispense with a good notebook, pencil and eraser, as well as comfortable field clothing, a cap, gaiters, and mosquito or bear repellent.

Chapter Challenge

How would you explain the theoretical foundations and importance of the study of animal behaviour to someone without a university education?

Challenge Objective

To develop the ability to explain the value, importance, and foundations of science in a simple and direct manner, using language comprehensible to those outside the academic environment is an important part of our social role. You should practice explaining, in a clear manner, the reasons, methods, and rationale behind your science. If people do not understand what we do and its significance, they will not value science.

References

1. Wilson EO (1975) Sociobiology: the new synthesis. Harvard University Press, Cambridge, p 697
2. Alcock J (2018) Animal Behavior: International Edition. English Edition | by Columbia University) Rubenstein, Dustin and Alcock, John. In fact, the 11th edition of "Animal Behavior: An Evolutionary Approach", Portuguese translation by ArtMed
3. Krebs JR, Davies NB (1997) The evolution of behavioural ecology. In: Krebs JR, Davies NB (Eds) Behavioral Ecology, An evolutionary approach. 4th edn. Blackwell, Oxford, pp 3–12

Chapter 2
Curiosity, Behaviour, and Their Basic Tools

Animals often surprise us. They can do extremely interesting things—sometimes simple and joyful, other times shocking. These behaviours catch our attention and spark our curiosity. And there are a handful of people in the world who become obsessed with trying to understand why an animal behaves in a certain way. What is its purpose, if any? Under what circumstances does it behave in a particular manner? Is there any advantage, and so on? These are the people who study Animal Behaviour, Ethology, and Behavioural Ecology.

For example, wouldn't you be intrigued if you saw a male dragonfly submerge a female in water for up to 20 min after mating? (Fig. 2.1).

In this study [1] (Fig. 2.1), we were very intrigued by this male behaviour of submerging the female in water and, after many observations, readings, and discussions with colleagues, we determined that this is a strategy to ensure paternity. In other words, the male does this to guarantee that the eggs the female will lay in his territory have been fertilised by him. Fascinating, isn't it?

Dragonflies are extremely interesting animals, with many territorial disputes, carnivorous habits, and can be very aggressive, despite their delicate appearance. Watch this video, for example:

https://youtu.be/uIWe5VNGObQ?si=RZLtCB5fQUAfCDDk

(All videos mentioned in the book are available on the YouTube channel—Kleber Del Claro. https://www.youtube.com/@kleberdelclaro9213).

See another example in Fig. 2.2.

Spiders, especially crab spiders (Thomisidae), can hide among the reproductive structures of plants and wait there until a floral visitor appears, which they can then prey upon. Have you ever seen this? Isn't it intriguing? A Mexican colleague [2] found it so curious that she kept observing and discovered that some bees can detect the presence of the spider. The study shows that the camouflage of spiders on flowers, to avoid detection by bees, depends on the spider's morphology, colour, observation distance, and position on the flower.

Do examples like these spark your curiosity? Do they move you? Do you feel motivated to study this kind of subject?

K. Del Claro, *Behavioral Ecology*, https://doi.org/10.1007/978-3-032-13988-7_2

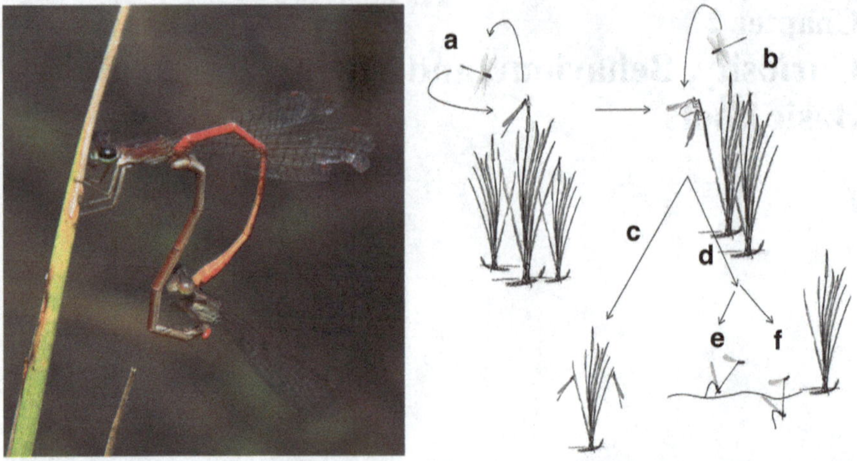

Fig. 2.1 Male *Telebasis carmesina* (Coenagrionidae), entirely red and clinging to a plant stem, holds the female's head with his abdominal appendages. The female places her genitalia next to the male's, on his second abdominal segment. On the right, a schematic drawing shows in "f" that males may submerge females in water after copulation

Fig. 2.2 A spider (Thomisidae) was hidden inside a flower in the Cerrado, right among the plant's reproductive organs, and when a small butterfly came to collect nectar, it became the spider's meal

If you feel unsure about answering these questions, it is worth knowing that a basic knowledge of zoology, botany, ecology, genetics, and evolution can be very helpful when deciding whether you will enjoy studying behaviour. It is important that you have strong reading skills and enjoy studying. You should like going into the field and be able to observe, with great patience and calm, the smallest details, often hidden by nature itself. See, for example, Fig. 2.3.

Fig. 2.3 A hummingbird clings to an inflorescence to collect floral nectar

I was only able to notice that this hummingbird was doing something unusual after observing it visit the inflorescences many times. At first, I didn't understand, because I had never seen a hummingbird hang from flowers to access nectar before. Is this common, and I just didn't know? Could it be an adaptation to the position of the flowers? If you are interested, go and study it! Go and read, see if the answer already exists in any publication, go ahead!

The images in this chapter are distinct examples of unique observational opportunities—simple, but only visible to those who go into the field with great patience. I know many people who, when in a natural environment, are only able to look at the landscape. They usually get bored quickly and say: "Right! We've seen everything here. Shall we move on?". Just imagine, if a person has no patience, no curiosity to look inside a flower, to listen to a faint sound, to breathe in the different scents, whether in the field or in the laboratory, will they be able to find something truly stimulating, something that could make them passionate about studying zoology, botany, or ecology? How will they spot something new if they lack the ability to observe with interest, curiosity, and a love for nature?

What I want to tell you is that there are countless stimulating examples and situations to study in behaviour. This is true all over the world: in tropical forests, tundra, deserts, and coral reefs, opportunities abound. But you have to enjoy what you do; you have to feel pleasure and happiness in doing it.

One thing that stands out in the cases discussed is that the studies reported and previously cited may have begun with a simple observation in nature or laboratory, noted in a small field notebook with a pencil, making corrections with a simple white eraser. This is another highly relevant factor in the study of animal behaviour. You do not need expensive or sophisticated equipment to become an excellent ethologist.

This is highly motivating, especially at the start of your career, or for an amateur who wishes to pursue this as a leisure activity. Some pertinent questions arise:

- What are the basic, essential tools and equipment required to study behaviour?
- How can you avoid making a significant financial investment in a type of study you are not yet certain you will enjoy?
- What are the initial precautions every researcher should take?
- How can you determine whether your chosen subject of study is of scientific interest?
- Am I on the right track? Whom should I ask for help, and when?

In this chapter and the next, I will try to help you find the path to these answers, to your initial questions, and to ever-increasing motivation. Here, we will begin with the tools, and in the next chapter, with the choice of subject, that is, the animal or topic of study.

2.1 Basic Tools: Your Field Backpack!

The tools I will present here will be discussed in other parts of this book, where I address their use, importance, and utility. Many of them you may already be familiar with; others vary in size, shape, and appearance depending on their application. I will focus primarily on the simplest, most affordable materials that are easy to obtain. In the notes on each item, I will comment on new developments, electronic equipment, and so on. Let us begin by listing what would be useful to have in our field backpack.

Notebook, Pencil, and Eraser—Yes, that's right, a notebook, pencil, and eraser. You might think I am mad, mentioning a notebook in the age of smartphones! But honestly, I know of no better place to make initial notes, and these should be made with a soft black pencil. This way, your notes will not be erased by water if your notebook gets rained on, falls into a puddle, or even a stream. Imagine your smartphone in an aquarium! In the tropics, torrential rain can fall at any moment, and no electronic device is entirely safe from the forces of nature. A sudden thermal inversion, common at the base of many European mountains, can simply render many electronic devices temporarily or permanently unusable, as can the scorching heat of hot deserts or the African, Brazilian, and Australian savannahs. Moreover, a thief is far more likely to be attracted to your expensive, attractive smartphone than to your humble field notebook. "That won't happen to me!" you might say. But it can! I know someone who had their work computer stolen twice in a row, with their entire doctoral thesis on it. Fortunately, the student had copies with me.

I am particularly fond of spiral-bound notebooks, as it is easier to remove pages with mistakes, turn the pages, and transcribe information into electronic spreadsheets in the laboratory. Clipboards with field sheets and tables can substitute for a notebook. For those working directly in water, a white PVC board can serve as a notebook, and you can write on it with an ordinary pencil as well. In some countries, such as the United States, you can buy waterproof notebooks that allow you to write in the rain. These initial tips are for all of us "mere mortals" who, at the start of our careers, cannot afford the marvellous equipment that can replace this basic, inexpensive material I recommend here.

Well, if you are financially well-off and know how to use electronic equipment, you may replace the field notebook with a tablet, laptop, or similar device. But be mindful of the elements and the risk of drops and falls. Always back up your data to an external hard drive, store your field data, tables, photos, and spreadsheets in the cloud: having copies is essential. New, fully charged batteries are, of course, basic equipment. Always make copies of your field data and keep them in different locations.

Portable Recorder or Smartphone—These can be very helpful, even occasionally replacing the field notebook, although there is the inconvenience of transcribing information and the risk of background noise interfering with your recordings. The recorder can also be very useful for capturing animal sounds. Nowadays, there are countless models of various sizes and prices. Choose the most accurate, with the longest battery life, smallest size, and quietest operation.

There are specific devices for recording the sounds of birds, amphibians, and aquatic animals, for example. Often, you will need to attach a directional microphone or a hydrophone (underwater microphone) to the device. These are considerably more expensive, so it is always wise to collaborate with colleagues who already have them, especially if you will not be using them continuously. Do not waste financial resources on unnecessary material goods or equipment you will not use extensively. Learn to share with your colleagues.

Camera/Camcorder—photography is both highly useful and enjoyable; the vast majority of the photographs in this book were taken by me. Some 20 years ago, a good camera with a set of macro lenses (for photographing small animals and details of flowers, for example) and telephoto (for animals and distant objects), was very expensive, and there were still the costs of film and its development. Nowadays, with the technological revolution of the last decade, everything has become much easier, but it remains very costly. Professional cameras are no longer absurdly expensive and can be easily found in most countries. The same applies to camcorders, which have also become much more compact, practical, and easy to handle. In fact, today, many smartphones have photographic and video capabilities that surpass those of some cameras and camcorders. A vast range of amateur cameras, with excellent resolution—many with zoom and others that allow good close-ups of flowers and small objects—can often replace professional cameras, as well as produce short films. This represents a true revolution in the documentation of behaviour, both animal and human. An important detail is that photos and videos can be stored, reconfigured, and electronically manipulated with the many computer programmes that emerge every day.

The multitude of features present in these new devices requires training for their use and handling. Reading the manuals is of fundamental importance. An excellent piece of equipment in the hands of an untrained person can yield frustrating results. A common occurrence is the user accidentally deleting images before saving them, due to inexperience. Computer programmes allow you to increase or decrease brightness, sharpness, etc. But be careful with their use so as not to artificially alter the reality of photographic or video documentation. Always preserve your credibility. Any alterations you make to photos must be mentioned in your work; this is a very serious ethical issue.

Binoculars/Hand Lenses—To observe animals at a distance, binoculars (or sometimes spotting scopes) are very important and useful. They come in all sizes, magnifications, and prices. If you are an ornithologist (or even an amateur birdwatcher) or a mammalogist, I recommend that you invest in a good one from the outset—it will be worth the cost. They can also be useful for entomologists, especially those studying butterflies, bees, and dragonflies. An additional advantage is that, if you give up everything related to behaviour, you can still use the equipment at sporting events, concerts, or to look at the moon, etc. It is not money wasted. Hand lenses are excellent for those working with small animals in the field, or for those who need to examine details of specific parts of captured animals or plants that are used or consumed by these animals. Those with built-in lighting and headbands are very good for those studying Pollination Biology. Once again, some smartphones can be useful, replacing hand lenses and allowing instant recording of the observed material.

Flashlight (Torch)—Another very useful piece of equipment, with a wide range of prices and sizes. Some can be used in both dry and wet environments, or even underwater. It is an essential item even for those who do not go into the field at night. You may need to illuminate a tree hollow, for example, or your car may break down, and you have to spend the night in the bush. For those working with invertebrates, having a piece of red cellophane to place over the light bulb is essential. Insects and spiders cannot see red light, and you will disturb the behaviour of these animals much less if you observe them under red light. It is also a basic safety item. Once, while working alone in the field at night—which is unwise—I realised I was being followed by someone who did not identify themselves. I walked a little towards the person, who moved away and hid. I attached my flashlight to a tree branch, pointing it towards the last place I saw the figure, and moved away in the opposite direction, circling through the bush until I reached the safety of my vehicle again. The flashlight was of fundamental help in this case. Remember to charge the equipment before going into the field, or to have spare batteries. More than one torch is essential for those working at night. We are diurnal animals; we cannot see in the dark, so we are highly dependent on this "extended phenotype."

Compass/GPS—The GPS (Global Positioning System) is now a basic tool for marking locations, measuring distances, and mapping natural areas. The good old compass is still very helpful, but GPS devices have become increasingly accessible as their prices have fallen. Today, smartphones already come with built-in GPS functions. These are easy-to-use, reliable devices in terms of the data they capture, store, and can transmit. For those working in the tropics, for example, in the Amazon, tropical savannas, the rainforests of South Asia, and in remote regions such as hot or cold deserts, the GPS is an essential part of any survival kit.

Information Collection Materials—Measuring tapes are used for measuring distances and heights (there are already electronic versions at an affordable cost). They are among the most useful pieces of equipment for recording fieldwork, followed by callipers, which allow the measurement of animals, body parts, and floral structures, for example. Stopwatches are indispensable for recording behaviour, to quantify the duration of behavioural acts, as are thermometers in all their variations, from the common to the more advanced, which record maximum and minimum temperatures and humidity. There are many types of "Data Logger"—devices that can be left in the field to measure temperature, humidity, and light intensity, and whose data can be transferred to your computer or smartphone. They are still expensive for researchers without funding, but are essential in some cases.

Measuring the oxygenation and pH of water may be necessary for ichthyologists, herpetologists, and limnologists in their studies associated with behaviour. The refractometer enables the measurement of sugar content (usually fructose) present in nectaries and plant exudates. Scales and spring balances (manual vertical balances) are essential when capturing small animals and determining their weight. In some countries, it is advisable to carry a statement regarding the use of spring balances to avoid being mistaken for a drug trafficker, as this equipment is frequently used for such purposes.

Do not be alarmed by the list; you will not need all these items. The size of your field backpack will depend on the questions you wish to answer. Some people will never use a pH meter, others will never use a refractometer, and so on.

General Collection Materials—In the field, especially if you are working in terrestrial environments and with invertebrates such as insects and spiders, certain materials are essential (Fig. 2.4):

- Plastic or glass jars of various sizes for storing and collecting insects, spiders, and flowers;
- Forceps and gloves to assist with the collection and/or handling of animals and plant structures in the field;
- An insect net or entomological sweep net for capturing insects;
- Tape, coloured string, and twine for marking plants, labelling, and identifying plant branches;
- Clipboard and/or field notebook;
- Pocket magnifier—if you have a watchmaker's loupe (Fig. 2.4), these are excellent for observing small invertebrates and flowers;

Fig. 2.4 A set of extremely useful field materials for the biologist's work, whether in animal behaviour, botany, entomology, or similar studies

- Plant press and pruning shears for collecting plant samples;
- A lightweight, collapsible stool, such as those used by anglers, if necessary;
- Toilet paper, which can be very useful, as well as adhesive tape and a wad of clean cotton wool;
- And, if you have allergies, never forget the medication prescribed by your doctor.

For different types of studies and animals, there are different types of equipment, often specific to a particular group. Seek information online, in specialised publications, or from colleagues. However, in general, the equipment listed here is almost always necessary at some stage of the study. For example, gloves are also useful when handling mammals (especially rodents and bats) and birds, as are capture nets. These nets are only sold with official state authorisation in many countries, especially in the tropics. Scissors may be required both for cutting small pieces of plant material and for preparing identification materials in the field. Often, you will need to know the plant species on which you observed your study model doing something—perching, feeding, resting, etc. You will then need to know how to collect and press the material for future analysis. Therefore, it is useful to know how to assemble a field press and how to properly prepare herbarium specimens. A simple field press consists of two wooden boards sandwiching two zinc plates (optional), sheets of cardboard, and then newspaper, with the plant material in the middle (Fig. 2.4). The whole assembly is tied together with cords or string. There are forceps of various sizes and shapes. Those in your backpack should meet your specific needs. For example, fine watchmaker's forceps for those working with very small animals or needing to manipulate flowers and buds; medium forceps with blunt edges for capturing small insects and spiders without injuring them; large forceps over 30 cm for handling dangerous animals such as spiders, scorpions, and bats. The researcher's good judgment should always prevail. An insect net (Fig. 2.4), made of tulle or organza, is a basic piece of field equipment for every entomologist. With it, you can capture butterflies, bees, wasps, and dragonflies to meet the needs of your study: sexing (examining the sex), marking the animal, measuring, and weighing.

I am not suggesting you take a field trunk, a whole removal van! Use common sense! For each field visit, there are essential materials, optional materials, and those for special situations. Good judgment should always prevail.

2.2 How Should I Dress for Fieldwork?

Clothing is one of the most important items for those working with behaviour. Yet few people pay attention to this. Clothing can interfere with the behaviour of the animal you are studying!

Field clothing should be comfortable; neither too warm nor too cold. It should protect exposed areas from excessive sun and from animals and plants that may cause injury, such as ants, spiders, scorpions, mosquitoes, and stinging plants. I always suggest covering the trouser cuffs with your socks or securing them with an elastic band. Think like a scorpion. It is on the ground, looking at a trainer with a trouser leg above it: "A little mound to climb up and enter a dark, damp, and cosy tunnel—what a delight!". This actually happened to Professor Helena Maura, my colleague, who fortunately was not bitten. We were more than 100 km from any hospital. Also consider mosquitoes, with so many viral diseases—dengue, yellow fever, chikungunya... A viral infection during fieldwork can ruin all your planning. Cover up, wear long sleeves, especially in the late afternoon.

A brief parenthesis here: it's important to note that if you are traveling to a region where diseases occur for which vaccines are available, such as yellow fever or dengue, make sure to get vaccinated in advance. Additionally, verify whether the country requires an entry visa and proof of vaccinations.

The color of your clothing is also very important, as we will see later, since it can influence animal behaviour by either stimulating or inhibiting certain responses. A red shirt in the field makes you highly visible and may scare animals away, for example. Footwear is also crucial. Consider whether you will be walking long distances, how long you will need to stand, or if you will be stationary in the laboratory; notice how your shoes or clothing affect your blood circulation. Boots and gaiters are important protection against snakes. Buy the best you can afford, preferably leather. In some regions of Brazil, jararaca snakes are extremely common. Mentholated talcum powder in your boots helps to repel ticks.

Hats, caps, and even a parasol may be necessary, depending on the degree of sun exposure you will face during your study. If you are working in the tropics or deserts, it is wise to budget for plenty of sunscreen (Fig. 2.5).

Fig. 2.5 One of my students was properly equipped for fieldwork under the scorching summer sun in the Cerrado (the Brazilian tropical savanna). Note the head protection, gaiters, boots, and the non-bright coloured shirt. She used sunscreen on her arms and face. When I took this photo (09:00 h, the temperature was 33°C and the air humidity was 17%)

Those working in aquatic environments will probably need Neoprene wetsuits. Opt for dark blue or black (which have less impact on fish behaviour), with at least 4 mm thickness for tropical streams, as they are quite cold.

2.3 Basic Equipment for an Animal Behaviour Study Laboratory

In the laboratory, there is a wide range of materials that may be useful, from chemicals to small household utensils. Ethanol, diethyl ether, hydrogen peroxide, acetone, formalin, agar-agar, sugar, and detergent are commonly used in behavioural studies, mainly for storing and preserving collected materials. Glass and/or plastic containers are also important, ranging from tiny *Eppendorf* tubes to specimen jars for urine and faeces analysis. Even large plastic or cloth bags can be useful for storing and transporting animals from the field to the laboratory. Once again, common sense should prevail. As I mostly collect small arthropods, I prefer transparent acrylic plastic jars with snap-on lids. They are very durable, lightweight, easy to handle, and allow for detailed examination of the animal inside. Often, it is possible to photograph body parts through the jar.

Petri dishes or plates, plastic trays, glass slides, hydrophilic and hydrophobic cotton, entomological pins, labels, tracing paper, filter paper, polystyrene plates, and pens for writing on plastic, glass, or metal are basic items that always come to mind when setting up a laboratory or purchasing consumables. Some equipment can be essential for quality work. Having a good stereomicroscope (bench magnifier) is fundamentally important for observing details of animal body parts or the behaviour of small animals such as ants and pseudoscorpions. This equipment, like the optical microscope, is indispensable for the identification phase of animal and plant material. To identify the species you are working with, or any of interest to your study, you will need to use identification keys that range from orders and families to genera, species, and sometimes subspecies. The structures distinguishing animals can be very small: a detail on a beetle's wing, a different hair on the back of a tiny bee. These pieces of equipment can be very expensive; researchers usually acquire them with support from government research funding agencies. If you do not have access to such equipment but need it, ask colleagues in a laboratory that does. Offer collaboration—one good turn deserves another.

It is always useful to have a precision analytical balance with at least three decimal places. Even for those working with bears, there may be a need to weigh and quantify small grass seeds occasionally found in their faeces. For small fish, amphibians, birds, and arthropods, minimal differences in animal weight can provide important behavioural information. There are a few other pieces of equipment I always like to have on hand, as they can serve multiple purposes. An aquarium, besides being used to house fish, can easily be converted into a terrarium. They are inexpensive—do not be tempted to pay a lot, opt for "do it yourself." Life should be fun. After all, what is an aquarium but five glass panes and a tube of silicone glue, assembled with patience and care? The most important thing is how you set up the aquarium. There are many tips on this in magazines, aquarium books, and YouTube channels. I have always assembled very simple aquariums that require almost no maintenance. The same aquarium, empty, with dry leaves, some stones and branches, and a little fine sand at the bottom, can be transformed into a terrarium for lizards and spiders, for example. It can also serve as an excellent micro-studio for photographing insects and other animals that would otherwise escape in the field.

Many other tools and pieces of equipment could be described here as basic instruments for the study of animal behaviour. As already mentioned, use your common sense—do not acquire all the materials at once. Assess what you will need first, and seek advice from more experienced colleagues. However, there is one tool without which you will achieve nothing. It is called **curiosity**!

Being curious is fundamental to becoming a good biologist, an excellent ethologist, and a first-rate behavioural ecologist. Cultivate this ability—curiosity can be developed. How? Read more, engage in more conversations with colleagues, go out into the field, visit a park or a flowering garden, close your eyes and try to listen, cover your ears and try to see further, try to detect the scent of flowers from a distance. Step away from the television, your beloved computer, and resist the lure of chat pages. Have real, face-to-face conversations with friends and pay attention to movements, details, different smiles, and nuances. You need to be a curious person, attentive to detail, and patient. This will even enhance your human perceptions.

Chapter Challenge

Imagine an animal you would very much like to study

Now make a list of everything you believe you might need to access or acquire in order to study this animal. Then separate the *ideal* from the *necessary* and adjust everything to what is *possible*!

Challenge Objective

Planning is essential for your study to yield positive results. It is very common for something to be forgotten at the crucial moment in the field. Therefore, the exercise of imagining a plan is enriching and will sharpen your mind, preparing you for a real future situation.

Go Further

In "The Science We Do," at the *link* below, you can download my latest book on behaviour in Portuguese or Spanish free of charge. There, you will find a chapter similar to this one, with more detail on equipment and methods for animal marking. Supplement your studies. If you're interested, you can also visit our laboratory and watch videos in English, Portuguese, and some in Spanish and Italian.

https://www.cienciaquenosfazemos.org/nos-team

References

1. Alves-Martins F, Del-Claro K, Jacobucci GB (2012) Sexual size dimorphism, mating system and seasonality of a Neotropical damselfly, *Telebasis carmesina* (Coenagrionidae). Int J Odonatol 15(4):263–273. https://doi.org/10.1080/13887890.2012.719422
2. Rodríguez-Morales D, Rico-Gray V, García-Franco JG et al (2018) Context-dependent crypsis: a prey's perspective of a colour polymorphic predator. Sci Nat 105:36. https://doi.org/10.1007/s00114-018-1562-0

Chapter 3
What to Study in Animal Behaviour?

Gathering field equipment is a delight. Packing your bag, grabbing the insect net, binoculars, water bottle—everything is wonderful. But then those questions arise:

- All right, I'm in the field—what should I study? Why should I study this?

Well, colleague, in Animal Behaviour we can follow two paths that are not mutually exclusive. Take a look at the marvelous jacamar in Fig. 3.1.

What a beautiful animal, isn't it? A long, sharp bill, very colourful plumage, an iridescent blue-green back, and it can reach up to 20 cm in length from head to tail. The animal is so striking that it captivated you, and you decided to study something related to it. Now, here are the two paths we can take.

The most common approach is to study the animal's behaviour and natural history and answer questions such as:

- What is its feeding behaviour? Is it insectivorous, as its bill suggests?
- What are its activity patterns? Is it diurnal, given its bright colours?
- What about its reproductive behaviour? Is the female similar to the male?

These are all questions directly related to your study subject, the focal animal of interest. But you can also ask different questions, more closely linked to a particular topic in Animal Behaviour, using the jacamar as the model organism for your research. For example, you might want to study territoriality in Galbulidae, the family to which this bird belongs, and begin your research with this particular species. Alternatively, you could investigate sexual selection based on plumage colouration. You might ask, for instance, whether the most colourful males achieve greater reproductive success, and compare your findings with studies in other birds or even other animal groups. Indeed, you can combine the study of specific individual behaviour with a broader topic of interest, such as sexual selection. This approach generally makes a significant contribution to scientific progress by clarifying broader and more far-reaching hypotheses. Imagine comparing the results of a study on sexual selection involving colouration, using this jacamar as a model, with a phylogenetically distant animal, such as a dragonfly (Fig. 3.2).

K. Del Claro, *Behavioral Ecology*, https://doi.org/10.1007/978-3-032-13988-7_3

Fig. 3.1 Male rufous-tailed jacamar (*Galbula ruficauda*)

Fig. 3.2 Male *Zenithoptera lanei* (Anisoptera: Libellulidae) in the Cerrado

In male odonates, both body size and fat content are associated with territory defence and reproductive success. Larger males with greater energy reserves win territorial disputes more frequently and attract more females. Wing colour is also considered a mechanism influencing agonistic behaviour among males, as wing pigmentation can serve as a signal of male quality. We analysed whether a set of physical (body size and wing colour), physiological (body fat content), and behavioural (male–male contests) traits in males of the Neotropical dragonfly *Zenithoptera lanei* (Fig. 3.2) were involved in territory defence and reproductive behaviour [1]. The results showed that wing colouration is an indicator of male quality, as this trait is related to resource availability and energy allocation. In the case of a male displaying intense wing colouration, together with other traits (e.g., large size, higher fat content, frequent patrolling), it can be assumed that this colouration is used to signal greater competitive ability to other males. This would discourage lower-quality males from invading the territory. Does the same occur in the jacamar you have chosen to study?

See the video about the study with *Z. lanei.*

https://youtu.be/_FVaa7oxxb8?si=sC6eG2dIzGHLDhHf

In another study we recently published [2], we showed that plumage coloration in birds can indicate whether or not they are diseased. Carotenoids are pigments obtained from the diet that determine plumage colouration in many bird species. In addition to plumage colour, they play a role in the immune response of many animals. In our study, we determined whether carotenoid-based colourimetric descriptors of plumage predict the occurrence of malaria parasite infections (order Haemosporida) in birds. To do this, we captured birds in a *strictu sensu* Cerrado area in southeastern Brazil, used PCR to detect haemosporidian infections at the molecular level, and microscopy

Fig. 3.3 A Rufous-collared Sparrow—*Coryphospingus cucullatus* (Thraupidae)—from the Cerrado. The purity and intensity of carotenoids, and the yellow, orange, or reddish colouration are negatively associated with the presence of parasites

to confirm all infections. We analysed the feathers of 58 individuals from three bird species: *Elaenia chiriquensis, Myiothlypis flaveola* and *Coryphospingus cucullatus*. The results of this study showed that colour saturation and carotenoid chroma enabled us to predict the individual occurrence of malaria parasite infections in these three bird species. Chroma, that is, the purity and intensity of carotenoids, was negatively associated with the occurrence of parasites. This means that the more yellow, orange, or reddish the bird, the healthier it was, and it could also display its best condition through more active behaviour, more frequent singing, as well as the intensity and brightness of its plumage colours (Fig. 3.3). Would this type of study not be able to address additional questions of sexual selection if you decided to investigate this aspect in the Rufous-collared Sparrow?

Can you see, then, that in the study of animal behaviour we can combine direct observations from ethology and natural history with broader and more theoretical questions?

We can study the sexual behaviour of the Rufous-collared Sparrow and relate it to higher body fat content (physiology), more colourful plumage (spectrophotometry), and a better immune response, resulting in greater individual fitness, which can be measured by offspring production. Isn't that fascinating? So, you can work with a species with basic ethological traits and natural history, or with broader topics. You can test theories, hypotheses, and look for patterns based on behaviour within phylogenetic lineages.

The study of animal behavior is a boundless world for knowledge.

And knowledge is a very important word. One of the main points in the study of animal behaviour is the understanding that you need to know and be familiar with your study subject, with the animal you are going to work with. Of course, you must know the specific names of each part of the animal's body; otherwise, how will you describe a behavior it exhibits?

3.1 You Must Know the Animal You Are Going to Study

Imagine you are in the field with a lecturer who studies pollination and florivory, and she turns to you and says:

"Look at the scopa on that bee!"

—Chirp, chirp, chirp... That awkward silence if you don't know what a "scopa" is.

See! This is a basic term in entomology. Certainly, the scopa the lecturer is referring to is a part of the bee. Then, the lecturer adds:

"See how it's full of oil!"

Oh dear, does that make it worse for you? Then look at Fig. 3.4.

By analyzing Fig. 3.4, you now know what a scopa is. Knowing the names of the parts of the animals you are going to study is fundamental. In the first study, Maura

Fig. 3.4 Centridini bee, *Centris* (*Ptilotopus*) sp, on a Malpighiaceae flower, *Byrsonima intermedia*. Note in detail the scopa (the region of the third pair of legs, covered with hairs) full of oil extracted from the glands of the plant's flower sepals

(Profa. Helena Maura Torezan-Silingardi, my main study partner) and I did together, we had to learn all these names [3]. We went into the field to look for a study for an undergraduate course and saw bumblebees performing buzz pollination on Myrtaceae flowers, which have open anthers and therefore exposed pollen. We wondered why this behaviour occurred. So, we had to learn about the bodies of bees, their capabilities, and the possible functions of each part. In the end, we concluded that this was a strategy to collect the remaining pollen from the plants after swarms of *Apis mellifera*, invasive bees that do not pollinate these plants, had visited. In this study, we combined animal behaviour with ecology, botany, and conservation. Note that we also had to learn the names of flower parts, types of anthers, and much more.

Thus, knowing your study animal and the universe in which it lives is essential for conducting a well-executed study.

The tips I most emphasize to help you gain self-confidence to study an animal in the field or laboratory are:

- study the names of the body parts of your chosen animal;
- read extensively about the animal; all aspects are important;
- talk to colleagues who have already worked with the animal or with a similar species;
- observe the animal thoroughly before starting your study; familiarise yourself with it;
- understand how the sensory organs of your model animal function: vision, olfaction, etc.;
- find out the animal's activity period;

- know where it lives and what it eats;
- try to learn the differences between the various life stages of the animal or between sexes;
- assess yourself and see if you feel confident to study this animal in nature.

These are some basic tips that, as you will see later, are crucial for achieving positive outcomes in your study and for ensuring success. From the previous tips, it is clear that simply knowing the morphological and anatomical aspects of the study animal is not enough. Understanding the sensory capabilities of your study organism is essential for obtaining robust research results. Let us consider, for example, Fig. 3.5. It depicts a bird, a mammal, and an amphibian. One flies, one runs, and one jumps, preferably always near water.

- Do these animals have the same auditory abilities?
- Do they possess the same visual capacity and the same ability to distinguish colours and chroma?
- Do they have the same olfactory capabilities?

Any biologist with a minimum of common sense would answer "no" to all three questions. A bird does not hear like a dog, which does not see like a frog, which does not detect odours like a bird. These animals have vast differences in these abilities. For example, some birds can see ultraviolet light, which enables them to locate food sources and distinguish between individuals of the same species based on feather patterns invisible to humans, dogs, and frogs, as far as we know. Dogs have about 300 million olfactory cells in their noses, compared to around 5 million in humans. They also possess an olfactory area in the brain about 40 times larger than that of humans, meaning they can process olfactory information in much greater detail. This is why dogs are often used in activities such as detecting drugs, explosives, and missing persons. Recent studies show that dogs can detect diseases and medical conditions, such as diabetes and cancer, through scent (e.g., [4]). Some species of tree frogs, such

Fig. 3.5 A Streamer-tailed Tyrant, *Gubernetes yetapa* (**a**), a savanna Fox, *Cerdocyon thous* (**b**), and a tree frog, *Hypsisboa* sp. (**c**)

as those in the family Centrolenidae, have an exceptionally high auditory capacity and can detect high-frequency sounds around 100 kHz, which is far above the human hearing range (about 20 kHz). This is due to the presence of a specialised middle ear, which acts as a kind of amplifier for high-frequency sounds. It is clear that the sensory world of animals is very different, both among themselves and in relation to our own sensory world. As observers of nature, we must respect and understand these boundaries. We cannot attribute actions or decisions to animals based on our own perceptions, unless we use appropriate equipment, careful and detailed observations, and field and laboratory experiments. If you do not recognise or respect these limitations, you may make serious errors in observation, recording, and behavioural analysis that compromise not only your study, but also have serious implications for species conservation due to inaccurate reports. Teaching with examples is much more engaging, so let us illustrate all this. See Fig. 3.6.

A colleague (Professor Vanessa Stefani Sul Moreira) conducted an extensive study on reproductive behaviour, parental care, and sexual selection in a very common spider in the cerrado of the Triângulo Mineiro region, the funnel-web spider, *Aglaoctenus lagotis* (Araneae, Lycosidae, Fig. 3.6). Our colleague and her collaborators [5] observed that males avoided the webs of gravid females, as these females would not accept further copulations, although they would respond to male signalling to attract them and subsequently attempt to prey on them. But how do experienced males distinguish the webs of gravid females from those of virgins? Lycosid spiders primarily use vision and proprioception to orient themselves in their environment. Vision is used to locate prey, mates, and obstacles, while proprioception allows the spider to sense the position of its legs and body to avoid collisions. Some species also use chemical cues on the web to orient themselves and find mates.

However, despite their abundance and diversity, Neotropical spiders have been the subject of relatively few behavioural ecology studies. Seeking to understand sexual

Fig. 3.6 Female *Aglaoctenus lagotis* (Araneae, Lycosidae) in her web

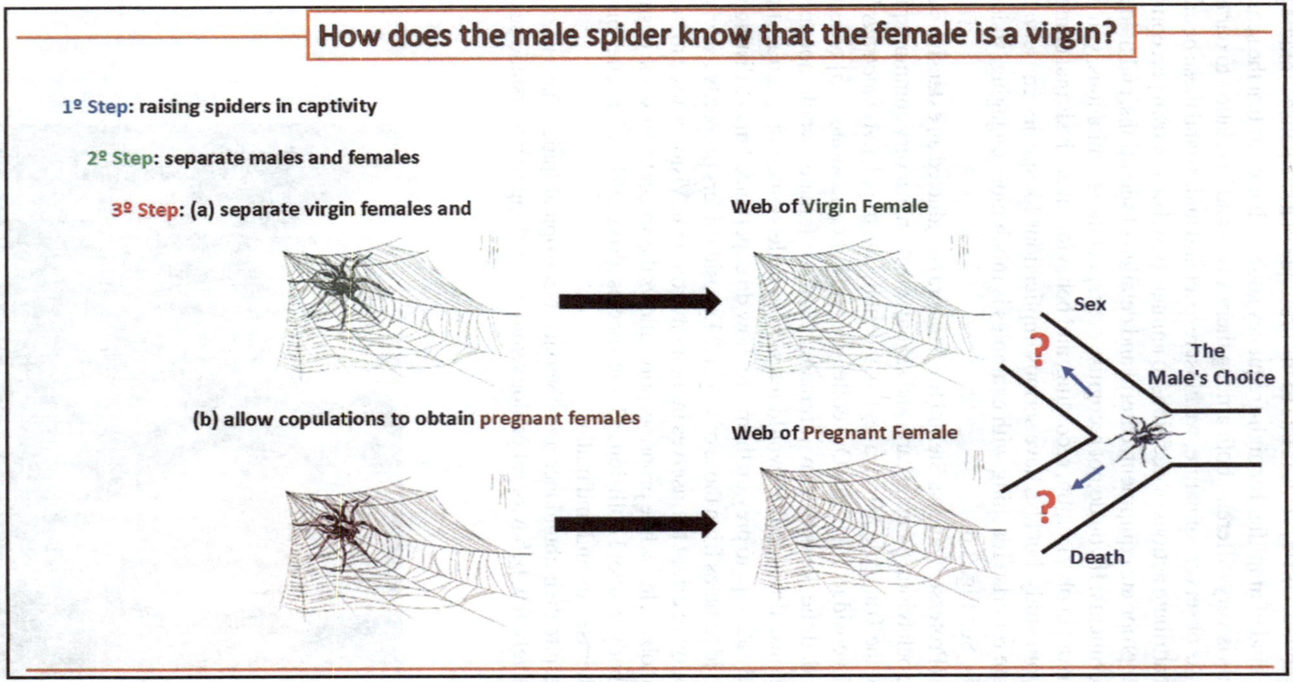

Fig. 3.7 Experimental manipulation: classic Y-maze experiment to test whether males of the funnel-web spider, *Aglaoctenus lagotis*, use chemical cues to decide which female to court. In this experiment (see Stefani et al. [5] for details), spider egg sacs were collected in the field and brought to the laboratory, where, after the juveniles emerged, they were separated and reared to adulthood. Then, webs from virgin and gravid females were separated and offered to the males for choice tests

behaviour and maternal care in this spider, Professor Vanessa and her colleagues [5] conducted experimental manipulations to test the hypothesis that:

> "males of this funnel-web spider are attracted by chemical, olfactory cues left by females in the web."

Two predictions were tested: (a) webs used by a virgin female will attract more males than those used by an inseminated (gravid) female; (b) the main cue in the web is chemical, so webs recently used by virgin females will attract more males than those not recently used. To carry out the experimental manipulations (Fig. 3.7), our colleague reared spiders in captivity to ensure she was working with virgin or inseminated individuals (allowing some to mate). She then allowed these spiders to construct webs. Next, the scientist carefully removed the females from their webs and offered the webs to sexually mature males in a classic Y-tube choice test experiment. In these experiments, the animal has two options, right or left, and a set amount of time to decide. The results showed that males were significantly more attracted to webs that had recently been used by virgin females. In other words, the perception was not visual, but chemical, based on the odours left by the females on the webs (Fig. 3.7).

Elegant studies such as this can open significant pathways in a scientific career. A publication like this in a renowned journal such as the *Journal of Natural History*, a British periodical established in 1841, gives you credibility, establishes your name in the scientific community, and opens doors. Therefore, if you truly wish to become a respected scientist in ethology, study your animal or research subject thoroughly before starting your work, conduct preliminary observations, read extensively, and consult with more experienced colleagues: self-reliance is essential. In this context, we have another important topic to discuss.

3.2 Individual Limitations

Imagine that you are someone who is very afraid of being stung by wasps or bees. You might even be allergic to their venom, putting your life at risk if stung. Picture yourself in this situation and then deciding to work with a large and beautiful wasp like the one in Fig. 3.8.

Do you think you could conduct a good study with wasps? In the journal Sociobiology [6], we published a study in which we assessed the ability of wasps to be more aggressive organisms and better plant defenders than ants. Many wasps, ants, and even spiders are attracted to plants by the offer of floral or extrafloral nectar, and these predators can protect plants against their natural enemies. In this study, we had to capture many wasps, handle them, mark and release them, or collect and mount them for identification. The chance of being stung—or rather, the near certainty that it could happen—was present on every field trip. So I ask you: if you had an aversion to wasps, would you be able to work with them? Would you be able to do a good job?

Many stubborn people, who would very likely get themselves into trouble, would say yes and claim:

Fig. 3.8 Wasp *Editha magnifica* (Perty 1834) (Hymenoptera, Sphecidae)

"Ahhh, if I want to, I can handle it, professor, you don't know me!"

Uh-huh! Look, working against your own limitations is a huge mistake. Everyone has their limitations. I have a colleague who works with all kinds of snakes, wades into the wetlands of Brazil and pulls anacondas out of the mud single-handedly. That would be a terrifying situation for most people I know. But this same woman simply faints if a butterfly caterpillar falls on her arm, because she has a huge phobia of that stage of butterflies and moths.

It is not just the type of animal—such as spiders, snakes, bats, and ants—that can be a limitation for some people. The environment and even the time of day can also be complicating factors, a limiting factor. Think carefully: if you cannot swim, would you work at sea with dolphins? Beautiful, isn't it? But you need to learn to swim first. And if you are not tolerant of the motion of the sea? Working at night in the forest with bats, in the dark, in the cold, at a time when dangers are greater—such as the presence of unwanted people, active snakes and spiders, as well as swarms of mosquitoes—can you handle it? What you need to understand is that individual limitations weaken and undermine your capacity for observation, and this will compromise your study. We have a pact with the truth; science depends on the truth of facts, and a person who is uncomfortable, inhibited, and unfamiliar with their working environment will not be able to uncover the truths of nature.

Someone with lower back problems will not be able to carry out a study that requires excessive time sitting; a person with special needs will require adaptations and assistance and will need to choose their study subject and location very carefully. It is not a matter of overcoming adversity, of being a super-scientist who fears nothing and overcomes all barriers. It is about being honest and recognizing your own limits.

Chapter Challenge

Imagine an animal you would like to work with but have never worked with before. Now draw it and label each of its parts with the correct names. Then list all the sensory abilities of this animal.

Challenge Objective
Knowing your subject is fundamental. Without understanding the morphology and sensory abilities of the animal you intend study, you will not be able to carry out a successful project. Compare the outcome of your challenge with the existing literature on the chosen animal and identify where you went wrong. Refine your skills!

Go Further
See this video that exemplifies this chapter's objectives:

https://youtu.be/_FVaa7oxxb8?si=sC6eG2dIzGHLDhHf

References

1. Ancco Valdivia FG, Alves-Silva E, Del-Claro K (2020) Differences in size and energy content affect the territorial status and mating success of a neotropical dragonfly. Austral Ecol 45:748–758. https://doi.org/10.1111/aec.12891
2. Penha VAS, Rodrigues R, Quaglia AI et al (2020) Plumage colouration predicts haemosporidian infection occurrence in birds. Ardea 108(1):39–48. https://doi.org/10.5253/arde.v108i1.a2
3. Torezan-Silingardi HM, Del-Claro K (1998) Behaviour of visitors and reproductive biology of *Campomanesia pubescens* (Myrtaceae) in cerrado vegetation. Ciência e Cultura 50:281–284
4. Rooney NJ, Morant SV, Guest CM (2019) Investigation into the value of trained glycaemia alert dogs to clients with type I diabetes. PLoS One 14(5):e0216715. https://doi.org/10.1371/journal.pone.0216715
5. Stefani V, Del-Claro K, Silva LA et al (2011) Mating behaviour and maternal care in the tropical savanna funnel-web spider *Aglaoctenus lagotis* Holmberg (Araneae: Lycosidae). J Nat Hist 45(17-18):1119–1129. https://doi.org/10.1080/00222933.2011.552802
6. Del-Claro K, Alves-Silva E (2016) Wasps are better plant-guards than ants in the extrafloral nectaried shrub *Ouratea spectabilis* (Ochnaceae). Sociobiology 63(1):705–711. https://doi.org/10.13102/sociobiology.v63i1.908

Chapter 4
Animal Behaviour: Basic Terms and Definitions

As with any major field within an evolving science, such as Evolutionary Biology, the definition of the scope of animal behaviour studies has also changed, adapted, and improved over time. For example, Nico Tinbergen defined animal behaviour as "*the set of movements performed by an animal in response to internal or external stimuli*" and emphasised the importance of studying it in its natural context. Edward Wilson went further and defined animal behaviour as "*coordinated responses of living organisms to internal and external stimuli, including the mechanisms of learning, memory, perception, and communication.*" These definitions highlight the importance of studying animal behaviour as a complex and dynamic phenomenon, involving multiple levels of organisation, from individual neurons to entire social groups. They also emphasise the need to understand the ecological, evolutionary, and genetic factors that shape animal behaviour. Perhaps we can simply define it as *everything an animal does or... refrains from doing*.

To illustrate: when we see an animal running, jumping, grooming itself, parasitizing, or preying on another, it is clear that the animal is exhibiting some type of behaviour (Fig. 4.1A).

In Fig. 4.1A, we can clearly observe the behaviour of the snake, swallowing the small bird head first. In the field, in a real situation, we can note every detail of this behavioural sequence, from approach to capture, from constriction to ingestion of the prey by the snake. But from Fig. 4.1B, we know that the snake had recently fed on a parrot because we witnessed the process. However, another observer arriving after the predation event would only see a fat, motionless snake for a long period, which may last hours. And in this situation, is the snake doing nothing? It is important to remember that animals can display behaviours in which they refrain from activities involving movement or displacement. To human perception, it may seem as if they are doing nothing. For example, sleeping, hibernating, or feigning death, which we call tonic immobility or thanatosis. Even when an animal appears to be doing nothing, this "doing nothing" also represents a type of behaviour and serves a function. Thus,

K. Del Claro, *Behavioral Ecology*, https://doi.org/10.1007/978-3-032-13988-7_4

Fig. 4.1 In (**A**) we see a boa (*Boa constrictor*) preying on a maritaca in the Cerrado, and in (**B**), the same boa, some time later, completely motionless on the ground. Photos: José Henrique Pezzonia

we can understand behaviour as the sum of all acts an animal performs or refrains from performing.

In this book, we will combine theory with examples and basic practical concepts in the study of animal behaviour. I believe that the flexibility of thought, which can and should occur in scientific educational texts, reduces the monotony of reading that some people experience, distracting them and diminishing learning.

So, let us now move on to a bit of practice.

4.1 Basic Initial Terms in Animal Behaviour

I imagine you may have found Chapter 1 of the book a little tedious, as it contains a lot of history, and that now, as you begin this more practical section, you are eager to understand how to describe what an animal does or refrains from doing. So, let us begin this task by explaining that there are some basic terms that are fundamental for us to start a description. Behaviour is now a widely researched field, not only among biologists—including ecologists, zoologists, geneticists, parasitologists, and even some botanists—but also among psychologists, veterinarians, animal scientists, medical doctors, and some agronomists. Therefore, it is necessary from the outset to establish some standardisations for a better understanding of the study of animal behaviour and its applications in Behavioural Ecology. At many conferences in the field, and even in scientific journal articles I read or review or edit, I frequently encounter mistakes in the use of descriptive terms for animal behaviours. Let us look at some examples.

In Fig. 4.2, how would you describe the animal's posture and position? Observe, think, and try to write it down on a sheet of paper before continuing to read. Stop! Do not go any further, do as I ask—it is a good exercise!

Fig. 4.2 Plush-crested jay, *Cyanocorax chrysops*

When we speak of "posture," we are referring to *the arrangement of the animal's body parts in relation to itself*. Thus, the jay displays a perching posture, that is, it is supported on a substrate (usually a branch or trunk), with its wings closed, body perpendicular to the substrate ("upright" posture), head facing forward, and beak closed.

As for "position," we can say that the animal is supported, or perched, on a tree branch. In describing position, we now seek to indicate *the relationship between the animal's body parts and the habitat*, the nearest substrate, a reference point in the environment. Obviously, it is difficult to separate posture from position in behavioural descriptions. The simplest and generally most accurate way to describe the animal in the photo would be simply to say that "*the jay is perched on a tree trunk of 'tree'*," so the term "perched" refers to posture, and "on a trunk" to its position. This may seem trivial, but it is actually quite common to find, in scientific texts and conference posters, confusions such as: "the jay is positioned with its wings closed, in a frontal posture to a fruiting tree." This is a complete inversion of the correct use of basic descriptive behavioural terms. Using terms correctly, in any scientific field, is of fundamental importance, as standardisation allows for comparisons.

Let us now consider two other commonly used terms. Analyse the photos below (Fig. 4.3) and try to describe the animal's behaviour, not only in terms of posture and position, but also regarding the "movement" and "displacement" performed.

The word "movement" should be used in the sense of *describing changes in posture*, that is, of the animal's body parts in relation to itself. "Displacement" refers to *changes*

Fig. 4.3 Bee *Centris* (*Ptilotopus*) sp., Apidae, Centridinae, pollinating *Byrsonima intermedia* (Malpighiaceae) in the Cerrado

in position, that is, of the animal in relation to environmental substrates. In general, in behavioural descriptions, we use both terms together in the same description. In this example, we can say that in figure (A), the bee moves its wings synchronously forwards and backwards, while displacing itself towards the inflorescence. In figure (B), it clings to a flower and, with movements of its forelegs, collects pollen and deposits it in its scopa with movements of the midlegs, while lateral displacements of the body occur. These two terms, movement and displacement, are often used interchangeably far more frequently than you might imagine.

Some people consider it pedantic on our part when we insist on the correct use of ethological terms. Learn this: care in description, the use of correct language, and precision in scientific writing are what will increase the chances of your text being well understood, the success of your message transmission, and, ultimately, the likelihood of your work being accepted for publication. If someone, in the past, took the trouble to standardise scientific language, in any field of knowledge, we should, out of respect and responsibility, adopt and use this language, or question it and propose a new one. Such standardisations usually contribute positively to better understanding among researchers, to improved comprehension of scientific studies, and to avoiding the same behaviour, whether common or stereotyped, being described repeatedly by different researchers, taking up considerable editorial space and often causing great confusion. Let us give an example of a major advantage of standardising certain terms.

I will use a very simple and well-known term among us biologists: to hover. See my description: "The hummingbird *hovered* in front of a flower in the Cerrado." What does it mean to hover?

This term means to remain suspended, that is, to beat the wings in such a way as to fly without moving, as if stationary in front of a flower, as in this case (Fig. 4.4). Imagine if you had to describe this behaviour every time you wanted to say how a hummingbird was visiting the flower you were studying. This is a typical behaviour

Fig. 4.4 Hummingbird, *Eupetomena macroura* (Trochilidae) hovers in front of a *Qualea multiflora* (Vochysiaceae) flower collecting nectar

of hummingbirds, and the existence of a common and well-known term greatly facilitates matters, saves editorial space, and ensures that everyone knows exactly what you are referring to.

In the following section of the chapter, I will discuss "ethograms" (behavioural repertoires) and you will see the enormous advantage of standardising terms. I will also draw your attention to the advantages and challenges of working with animal behaviour, and the precautions we should take. I will try to show you how behaviour can be a useful tool not only in behavioural ecology studies, but also in other research, such as pollination, seed dispersal, interaction ecology, and so on.

4.2 Ethogram—Behavioural Repertoire of a Species

Ethograms or behavioural repertoires are very useful tools for understanding all the behavioural possibilities of a given animal and for guiding the planning of our studies. They are especially important and interesting in the initial stages of research, when the research question is still being defined, as they help to answer basic questions such as:

- What are the main activities of the animal under study?
- What is its peak activity period?
- What types of interactions does it exhibit?
- How does it allocate its time throughout the day?
- Are there behavioural differences between sexes or age groups?

You will now realize how important it is to be thoroughly familiar with the terms you will use to convey to your peers an accurate description of your observations.

It is through this simple method of behavioural observation and recording that many interesting questions and hypotheses arise, which can later be tested through experimental manipulation, as we will see in the following chapters. Ethograms are indeed an excellent way to generate ideas. To compile a behavioural repertoire for a species, that is, to describe everything it does or does not do, I recommend dividing your study into three stages.

In the first stage, you will conduct preliminary observations and a literature review. The main objective is to determine whether studies have already been conducted on the species in question and, among other things, to identify the peak activity times and locations for the species. You can, for example, record the number of behaviours exhibited by the animal or animals under study during the first ten minutes of each hour of the day. By repeating this three times on days with similar weather conditions (if it is a field study), you will be able to determine the periods of greatest activity for your study subject. A recent study providing a clear example of this approach is found in [1]. In this field study, knowing in advance that the subject was a diurnal ant, we observed from seven in the morning to seven in the evening how many individuals entered and exited the nests of the ant *Camponotus crassus* in the Cerrado. The aim was to confirm the species' peak activity period so that we could subsequently observe its foraging, nest defence, and other activities.

On the "The Science We Do" page, in the videos section, you can watch a short video in which I briefly explain how this study was conducted: https://www.cienciaquenosfazemos.org/videos

In the second stage, you will qualify all observed behaviours. Using the all-occurrences sampling method (which will be explained in the next chapter), record everything the animal does or does not do, and begin to categorise these *behavioural acts* into *types* and *patterns*. It is important to clarify that *type*, or *category*, as some colleagues prefer, refers to behavioural categories that can be subdivided into *patterns* and *acts*, with their component parts. For example, "flying" fits within the *locomotion type*, which includes: walking, running, swimming, flying, etc. The *pattern* relates to the type. In this case, the *behavioural act* would be termed "flying," "taking off," "landing," etc. The *behavioural act*, in turn, refers to the *pattern*. The description of each movement within this *pattern* characterises the component parts of the behaviour (see Table 4.1).

The number of sessions required at this stage will depend on the species in question and what is indicated by the literature and your own experience. For ants, a thorough qualification phase may take approximately 30 to 40 h. For this animal group, we know that we have a representative sample of behavioural acts when we have qualified more than 35 acts. This "magic number" is again determined by the literature and our years of data collection.

Once the qualification is complete, we move on to the **third stage**, which is the time to quantify, to record how many times in each session the animals display each of the observed behaviours. You will do this by following one of the observation and recording methods that will be taught in the next chapter. Be rigorous and disciplined with yourself at this stage.

Table 4.1 Behavioural repertoire of soldiers of the ant *Cephalotes pusillus* (Klug, 1824; Formicidae) under captive conditions ($n = 4$ colonies, 100 h of observation)

Behavioural categories	Colonies				
Behavioural acts	1	2	3	4	Total
Exploration	**24,090**	**16,800**	**15,440**	**5,780**	**9154**
Stationary outside the nest	10,850	2440	3110	1160	2170
Walking inside the nest	5330	11,490	8070	3020	4540
Walking (exploring) outside the nest	7910	2880	4250	1600	2000
Defense	**35,600**	**20,610**	**12,970**	**21,450**	**21,420**
Stationary at the nest entrance	35,090	20,530	12,710	21,440	21,410
Agitated at the nest entrance	0.510	0.074	0.260	0.010	0.080
Feeding	**4310**	**4062**	**8270**	**5390**	**5498**
Antennating termites	—	—	—	—	—
Trophallaxis with worker	2280	2510	3340	3750	3480
Trophallaxis with soldier	1120	1220	3470	1380	1570
Trophallaxis with queen	—	0.110	—	—	0.010
Feeding on diet (pulp)	—	0.140	0.840	0.260	0.290
Feeding on honey and water solution	0.910	0.074	0.620	0.005	0.140
Communication	**5730**	**6170**	**10,440**	**18,640**	**15,640**
Antennating worker	2790	2440	3696	6750	5730
Antennating soldier	1170	1180	3664	2240	2220
Antennating at nest entrance	1780	2550	3080	9640	7696
Cleaning	**1620**	**1630**	**1300**	**1060**	**1175**
Cleaning antenna (first pair of legs)	1370	1290	0.810	0.990	1027
Cleaning soldier	—	—	—	0.010	0.007
Carrying dead ant	0.250	0.330	0.490	0.050	0.140
Carrying waste	—	—	—	—	—
Resting	**28,650**	**50,740**	**51,590**	**47,680**	**47,040**
Soldier standing in the nest	28,650	50,740	51,590	47,680	47,040
Total	**100**	**100**	**100**	**100**	**100**

Other examples of studies on behavioural repertoires can be found for other ants, such as in [2] or [3, 4]. There are also examples involving arachnids, such as scorpions [5] or harvestmen [6].

However, one final question remains in this chapter. An example of the importance of an ethogram for the effective conduct of behavioural studies in the field. Let us consider an example.

Ants are very common in terrestrial ecosystems, especially in the Brazilian Cerrado savanna. They play various roles in the food web, including the protection of plants against herbivores. Although there are many studies on ant–plant interactions in the Cerrado, we still know little about the natural history of most ant species. An important example of an ant in the Cerrado is *Camponotus crassus*. This species is considered one of the main plant defenders in this savanna region. Nevertheless, there had been no specific studies on this ant, which hindered our understanding of how interactions between arthropods (such as ants) and plants are structured in the Cerrado. Therefore, in a study with [1], we provided a detailed description of the natural history and foraging ecology of the ant *Camponotus crassus*.

We conducted fieldwork over two years in different areas of the Cerrado, measuring the number, density, distribution, location, and internal structure of nests, the daily foraging behaviour of ants outside the nests, the food resources they collect, and whether there is territoriality and dominance by ants on the ground and on plants.

In this study on the biology and natural history of one of the most common yet least studied ants of the Cerrado, it was crucial to first establish a behavioural repertoire for the species. With this in place, it became much easier to define the fieldwork stages. By knowing the ant's activity schedule, we were able to map variations in their activities throughout the year and collect significant data on their feeding habits, movements, and interactions (Fig. 4.5). This enabled us to observe nests and their behaviour in hollow logs, on the ground, with different internal structures, and to discover that these ants forage for food during the day, both in the rainy and dry seasons.

In this example with *Camponotus*, note that the data on the number of individuals leaving or returning to the nest at each time of day were combined with temperature and humidity data (Fig. 4.5). We also took the opportunity to record the types of food brought back to the nest, such as prey, seeds, and liquids. We found that the most frequently collected resources were extrafloral nectar and exudates from trophobiont insects, which accounted for 83.33% of resources in the rainy season and 30% in the dry season. These data demonstrated that these ants interact significantly and empirically with plants and their visitors. Additionally, other behaviours and interactions with other ant species, animals, and plants were recorded during the intervals of nest activity observations. Thus, it was possible to show that *Camponotus crassus* is a dominant species, mainly on plants, but also forages on the ground. This is the first study to analyse in detail the natural history and foraging ecology of this ant. It is diurnal, aggressive, territorial, and its main foraging strategy is climbing plants. This information helps us to better understand how ants interact with the Cerrado environment and how they fulfil their role in plant protection. This study further shows that it is a combination of different pieces of information that results in a robust study of natural history and animal behaviour. It is through basic observations of behaviour, biology, and natural history that we are able to construct the life history of a species that is highly significant for the biodiversity of the Cerrado.

An important aspect within this small yet comprehensive universe of science is knowing how to properly record behavioural acts. Recording errors lead to mistakes in analysis and interpretation, which can seriously undermine our understanding and conservation of a species.

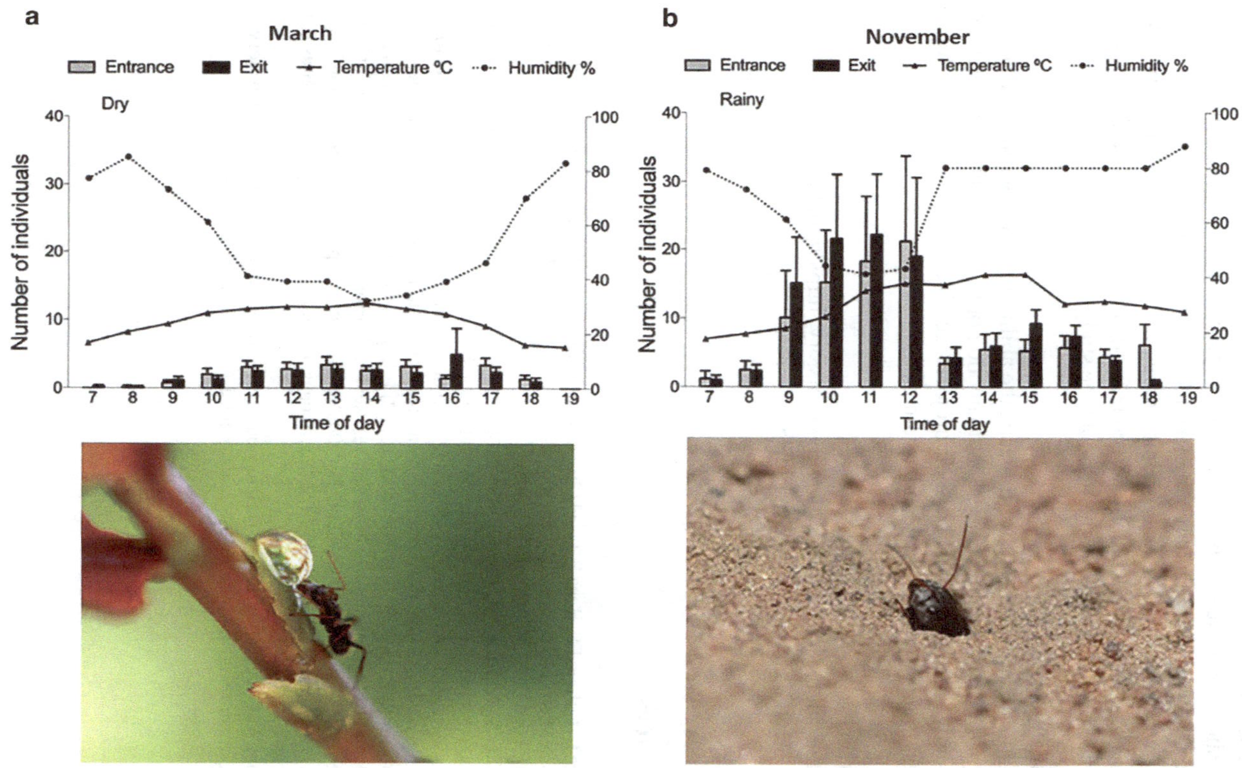

Fig. 4.5 Activity schedule of *Camponotus crassus* in Cerrado vegetation across different seasons

In the next chapter, we will discuss techniques for observing and recording behaviour.

Chapter Challenge

Write a short text of ten lines in which you explain to a colleague from another field the importance of standardising terminology in science.

Now write the same text, this time addressed to someone you would like to become a partner in this research; someone from your own field and therefore more experienced. Try to persuade them to join the study.

Challenge Objective

Training in scientific writing for projects, the ability to "sell your idea," is a crucial part of an academic career that many do not know how to do or cannot do competently, and for this reason, their projects are hindered or rejected by funding agencies. Practise your writing and argumentation skills.

Go Further

Watch this short video about *Camponotus* history: https://youtu.be/ALlLxeH2OCI?si=vHabt9ryq4KEjeHS

References

1. Lange D, Calixto ES, Rosa BB et al (2019) Natural history and foraging ecology of *Camponotus crassus* Mayr, 1862 (Hymenoptera: Formicidae). J Nat Hist 53(27–28):1737–1749. https://doi.org/10.1080/00222933.2019.1660430
2. D´ávilla S, Andrade FR, Prezoto F et al (2005) Activity schedule and foraging in *Cephalotes atratus* (Hymenoptera: Formicidae, Myrmicinae). Sociobiology 45(1):105–118
3. Barros-Miguel T, Del-Claro K (2005) Age polyethism and behavioural repertoire of *Ectatomma opaciventre* Roger, 1861 (Formicidae: Ponerinae). Revista Brasileira de Zoociências 7(1)
4. Del-Claro K, Santos JC, S JÚNIOR AD (2002) Ethogram of the arboreal ant Cephalotes pusillus (Klug, 1824) (Formicidae: Myrmicinae). Revista de Etologia (São Paulo), Brazil 4(1):31–40
5. Mineo MF, Assis GAF, Del-Claro K (2003) Behavioural repertoire of the yellow scorpion *Tityus serrulatus* Lutz & Mello 1922 (Scorpiones, Buthidae) in captivity. Revista Brasileira de Zoociências, Brazil 5:23–31
6. Pereira W, Campos A et al (2004) Behavioural repertoire and notes on the natural history of the harvestman *Ilhaia* sp. (Opiliones: Laniatores). J Arachnol USA 10:113–127

Chapter 5
Methods for Studying Animal Behaviour: Classical Tools

We now move into a more technical phase of the book. Since the study of animal behaviour began, various professionals—biologists, psychologists, naturalists, veterinarians, and medical doctors—have sought common methods and techniques that allow for standardisation and comparison between studies. This is fundamentally important, as the scientific method is based on objectivity, replicability, and verifiability. It requires the formulation of testable hypotheses and the execution of controlled experiments to evaluate these hypotheses. The results obtained are subject to peer review and must allow for verification and reproduction by other researchers. In short, the scientific method is an investigative process that seeks to discover and understand natural phenomena through empirical evidence and critical analysis, thereby contributing to the advancement of scientific knowledge. Hence, there is a need for standardisation in many fields of knowledge to ensure replicability and confirmation of observed phenomena.

Below, I will describe and provide examples of the four most frequently recommended behavioural sampling techniques: *all occurrences*, *sequences*, *instantaneous*, and *focal animal*. This standardisation follows the classic study by Altmann [1] (Fig. 5.1), a charismatic professor at Princeton University, USA, well known for her research on baboons and for defining classical terms that have been instrumental in standardising the study of animal behaviour.

It is particularly interesting to note that Professor Altmann began her career as a mathematician, later graduating in Biology and combining knowledge from both fields to help us better understand the behaviour of children and primates. But what are these terms she defined? How should they be used? Let us see.

K. Del Claro, *Behavioral Ecology*, https://doi.org/10.1007/978-3-032-13988-7_5

Fig. 5.1 Professor Jeanne Altmann honouring us with a visit to the poster of one of the studies we presented at the Animal Behavior Society conference, Princeton, USA, 2014. She kindly agreed to pose for this photo with me, which was a moment of great joy, and I felt honoured. We must always recognize those who paved the way of science for us

5.1 All Occurrences Sampling

In English, you will find the term *all occurrence sampling*, which actually refers to ad libitum behavioural sampling. In this case, you record everything you observe. Everything the animal does or does not do is of interest at this stage. This method is advantageous in various situations, for example, at the beginning of a project, during the familiarisation phase with your study subject. It is also useful for standardising the methodology you will use and for establishing intra- and inter-observer reliability.

I consider it fundamental during the qualification phase of behaviours for the development of a behavioural repertoire, or ethogram. This ad libitum sampling technique is also very useful for recording fortuitous, rare, or unexpected behaviours.

Here is a very simple example to illustrate. For instance, I had never seen a mule deer (Fig. 5.2). For me, everything this animal did would be new, unexpected, and unknown. I had never read anything about it. So, when I encountered one in Estes Park, Colorado, USA, if I wanted to record the behaviour of this beautiful animal, I would have to use all occurrences sampling.

Now imagine that I and another colleague, who also had no prior experience with this animal, had to describe its foraging behaviour—in this case, its search for food. The best approach initially would be for us to observe the animal independently, recording everything ad libitum. We would then sit down to discuss and compare our observations. This process should be repeated over several sessions until standardisation between observers is achieved, when the overlap of coincident observations exceeds 90%. This point would define the reliability of the observation between observers.

Fig. 5.2 Colorado mule deer, *Odocoileus hemionus*—a resting male

5.2 Sequence Sampling

This is the type of sampling in which the order of events is what matters. You are observing an event that occurs in stages, and every detail is important. Therefore, you cannot lose sight of the animal, which makes it difficult to record time, frequency, or the behaviour of other individuals if it is a group, so that even taking notes becomes complicated. If it is possible to use a recorder or a video camera, this is one of those situations where such equipment can make a significant and positive difference.

But in which situations should *sequence sampling* be used?

For example, describing an event related to reproductive behaviour, such as courtship and copulation. How does the male react upon perceiving the presence of the female, and what does he do next? How does the female signal to the male whether or not he is accepted for courtship? How does the approach, courtship, female submission, and finally copulation occur? How long does each of these phases last? Can or should they be subdivided? Some of these stages will certainly be subdivided into other, more detailed sequences that will require your full attention. In some situations, the same event may be observed several times, which will help you define the sampling sequences. In other cases, you may be dealing with an event that is not very frequent, or even rare, which will require you to recognise this and change your method quickly. In such cases, resort to sampling all occurrences, at least initially. In this type of situation, quantifying time with a stopwatch, recording changes in movement and displacement, positions, and postures may be necessary.

Let us look at a real example of a predation event in which the combination of sequence sampling and sampling of all occurrences was crucial to the success of the research. See Fig. 5.3.

Fig. 5.3 Wasp *Brachygastra lecheguana* opening flower buds of Malpighiaceae in the Brazilian Cerrado savanna to access and prey on Curculionidae beetle larvae. Photo: Torezan-Silingardi, H.M

A decade ago, Professor Helena Maura Torezan-Silingardi, working on pollination and florivory in Malpighiaceae of the Cerrado, observed that some species of wasps violated the flower buds and removed insect larvae from inside, which they then consumed. After a considerable period of preliminary observations, sampling all occurrences, and finally sequence sampling, our colleague was able to conclude that these wasps possess a remarkable ability: they can identify with almost 100% certainty which buds are parasitised by the endophytic herbivore [2]. Sequential sampling, and the understanding that there was a clear sequence in the wasps' behaviour, allowed this initial study to give rise to a series of further investigations. The most elegant of these from a Behavioural Ecology perspective was recently published in the historic and renowned journal *Behaviour* [3].

In summary, Mendes-Silva et al. [3] demonstrated that, in the Brazilian tropical savanna, the wasp *Brachygastra lecheguana*, of the subfamily Polistinae, is specialised in preying on certain endophytic beetle larvae that develop inside flower buds. Our group formulated and tested the hypothesis that these wasps use a combination of different mechanisms, such as visual, chemical (odour), and possibly tactile cues to locate the beetle larvae. Using a combination of experimental manipulations (visual, chemical, and visual/chemical), we tested the wasp's ability to detect endophytic larvae directly in the field. In addition, we assessed its ability to detect vibrations produced by beetles inside the flower buds. Our results, strongly based on behavioural sequence sampling, suggest that the wasp *B. lecheguana* employs a sequence of ecophysiological mechanisms to locate the endophytic larva within the flower buds: vision, olfaction, and possibly touch. The use of multiple cues by this wasp ensures such a high rate of predation on endophytic beetles that the wasp may have positive implications (reduction in beetle infestation) for the future reproduction of the host plant. Thus, you can clearly see the importance of sequence sampling in specific situations, particularly those involving predation, parasitism, reproduction, and parental care.

5.3 Instantaneous Sampling

The so-called *snapshots*, *instantaneous samplings*, or "behavioural act photographs" are widely used, especially for slow behaviours or when dealing with a very large group of individuals and aiming to sample the behaviour of a population. For example, have you ever considered recording the movement behaviour of a sea anemone? The impressive distance of 2 cm may take a week to cover. Another interesting situation is when you wish to record the behaviour of a large group of individuals, for example, a colony of ants, termites, bees, or pseudoscorpions. Yes, there are social pseudoscorpions and we study them (Fig. 5.4).

A colony of pseudoscorpions of the species *Paratemninoides nidificator* (Ateminidae) may consist of a few dozen to two hundred individuals. These colonies

Fig. 5.4 Colony of *Paratemnoides nidificator*, pseudoscorpions. In detail, we see females with their egg sacs and some adults, both males and females, nearby

include reproductive and non-reproductive females, nymphs of various ages, silk chambers for brood and moulting, and males. All perform some function simultaneously [4]. How can one record the behaviour of the colony and construct an ethogram? In this case, the best approach is to have a list of behaviours previously obtained through sampling all occurrences, and then make spaced, controlled observations, following instantaneous sampling and recording the number of individuals performing each behaviour from your list sequentially: a snapshot!

A detail of this type of sampling is that the time intervals in instantaneous sampling should be somewhat longer than those time intervals in an "all occurrences sampling." This is because, often, in instantaneous sampling, you work with behaviour lists and generally large groups of individuals, and sampling at very short intervals can overestimate longer behaviours and underestimate faster or rarer behaviours.

5.4 Focal Animal Sampling

Focal animal sampling should be used for animals or groups of animals that can be "easily" observed and identified. By "easily," I mean those groups of animals that allow the observer to approach closely and become accustomed to their presence. In this type of sampling, an individual from the group is observed at defined time intervals, and its behaviour is recorded at the moment of observation. When applied to a group, this technique can, in practice, closely resemble a sequence of "snapshots" taken one by one of each group member.

In this technique, an individual is systematically chosen at a time as the main subject of observation. Selection can be based on predetermined criteria such as age, sex, hierarchical position, or even at random. The behaviour of the focal animal is recorded in detail. This may include observations of actions, movements, social interactions, feeding patterns, resting, reproduction, and other relevant aspects. The observation period may vary depending on the study, ranging from minutes to hours or days. The frequency of observations can also be determined according to the study objectives. Remember that all animal observation should take into account the environment in which the animal lives, including factors such as climate, vegetation, the presence of other animals, and any element that may influence the behaviour observed.

This technique is widely used in the observation of mammals, such as primates (including humans!), and birds. The observation of social animals living in groups, where hierarchy is evident, often relies heavily on this type of observation. It can also be applied in captive conditions, such as in zoos. A reminder: whether studying animals or humans, you must always obtain the necessary approvals from the relevant ethics committees. University, state, and federal research ethics committees are your guarantee of avoiding legal issues in your study. **Preserving your name and academic credibility are top priorities**.

As you may have noticed from the descriptions of these four methods, which are the most commonly used in ethology (Table 5.1), there can be considerable overlap

Table 5.1 Summary of sampling types and their use in animal behaviour

Sampling type	Situations in which it may be used
All occurrences	At the start of a study; familiarisation phase with the animal; used for behaviour qualification, for rare or fortuitous behaviours, and for standardising inter-observer methodology.
Sequences	When the sequence of behaviour is of interest; when a behaviour can be divided into phases or stages.
Instantaneous	For studying very slow animals; for studying animals in large populations; for ethograms of social insects in the laboratory.
Focal animal	For groups where each member can be individually identified. In the case of animals that can be observed individually, with little difficulty, in the wild and/or in the laboratory.

between them. This is common. Therefore, make sure to clearly define the methodology used so that, in your results, you can clearly indicate which method was predominant in your study. Establishing a clear relationship between method and result is fundamental to research success.

I would like to draw your attention to the importance of reading. Published scientific articles, followed by theses, dissertations, and textbooks, will be your main sources of reference to guide you in choosing the best sampling method. With a solid grounding in the literature, you will benefit much more from discussions with your supervisor or colleagues in the field. Do not overlook good in-person scientific conferences, coffee breaks, and poster sessions—these are where things really happen.

Chapter Challenge

At home, on the street, or in one of your university's cafeterias (if you are a university student), choose an animal. It could be a cat, a dog, or a bird. For thirty minutes, record everything it does or does not do. Make a list. Now, for another thirty minutes, try to quantify these behavioural acts—how many times it performs each of these acts. Good exercise!

Challenge Objective

Observational training is essential for anyone wishing to study animal behaviour. Only through practice can you build self-confidence and gradually become independent.

Go Further

Read this paper: Altmann J (1974) *Behaviour*, Vol. 49, No. 3/4, pp. 227-267. https://www.jstor.org/stable/4533591 and get details about the life and work of professors Jeanne Altmann and John Alcock. Knowing who teaches us is a sign of respect for science.

References

1. Altmann J (1974) Behaviour 49(3/4):227–267. https://www.jstor.org/stable/4533591
2. Torezan-Silingardi HM (2011) Predatory behaviour of *Pachodynerus brevithorax* (Hymenoptera: Vespidae, Eumeninae) on endophytic herbivore beetles in the Brazilian tropical savanna. Sociobiology 57(1):181–185
3. Mendes-Silva I, Queiroga D et al (2021) Multiple cues guarantee successful predation by a Neotropical wasp. Behaviour 159(7):643–655. https://doi.org/10.1163/1568539X-bja10144
4. Del-Claro K, Tizo-Pedroso E (2009) Ecological and evolutionary pathways of social behaviour in Pseudoscorpions (Arachnida: Pseudoscorpiones). acta ethol 12:13–22. https://doi.org/10.1007/s10211-009-0052-y

Chapter 6
How to Plan a Research Project in Animal Behaviour

All of us, at the beginning of a scientific career, ask ourselves:

"How do I start a research project?"

Clearly, behind this question lies not only the scientific desire to conduct a study, but also the wish to carry out research that is engaging and interesting, that people will appreciate and discuss, and that will truly make a valuable contribution to science. It also needs to be something you enjoy, that you find rewarding to dedicate your time to, as time is precious and once gone, cannot be recovered—time well or poorly spent is gone for good.

Therefore, I recommend that you look for something you genuinely enjoy studying, something you believe is truly worthwhile. Consider this example from a much-appreciated former student (as most of them are!). Raoni Rosa Rodrigues went to the Amazon to carry out his undergraduate dissertation together with another outstanding former student, now Professor Lucélia Nobre Carvalho. They observed a dwarf cichlid, *Apistogramma hyppolitae* (Fig. 6.1), in small backwaters of Amazonian streams. As you can see in the figure, this small fish displays different body colourations.

The different colourations of this fish immediately raised a question: why does this cichlid display such varied colour patterns? From this question, a hypothesis emerged: the different colourations of *A. hyppolitae* are associated with different behavioural types [1, 2]. This study enabled us to understand that animal colouration serves multiple functions, and that fish, among vertebrates, are notable for their great diversity of colour patterns. We described the colour patterns of *Apistogramma hippolytae* and confirmed the hypothesis that these patterns are expressions of the animals' social and behavioural characteristics.

In summary, we characterized colour patterns associated with different behaviours (Fig. 6.1): resting (P1), feeding (P2), aggression (attack and escape) (P3), and sexual and agonistic display (P4). Colour changes take only a few seconds. In the natural environment of this cichlid—shallow areas with clear water—colour-based communication indicates individual motivational state, social status, and readiness to reproduce, possibly helping to minimise energy loss through unnecessary interactions.

K. Del Claro, *Behavioral Ecology*, https://doi.org/10.1007/978-3-032-13988-7_6

Fig. 6.1 Dwarf cichlid, *Apistogramma hyppolitae*, showing its four most commonly observed colouration types in backwaters and lagoons of the Amazon

And your question remains:

"After discovering an interesting topic, something exciting to study in the field, how can I tell if this is novel, a valuable contribution to science?"

The first step we should take after finding something interesting in the field or laboratory, such as the study of this cichlid, involve conducting a thorough literature review. You should consult the Web of Science, Google Scholar, Scopus, Scielo—every resource available—and investigate whether that topic or species has already been studied in the context of your interest.

Then you should read extensively, discuss with colleagues, and return to the field for confirmatory preliminary observations. Indeed, it is possible that what you observed is a very rare event, unlikely to occur again. How would you proceed with such a case? The literature will tell you whether you are "reinventing the wheel" and whether what you wish to study is already well known and established. If so, it makes little sense to persist with that topic.

6.1 Steps in the Study of Animal Behaviour

The first steps to follow in the study of animal behaviour are:

- Identify a good topic, an interesting question;
- Make preliminary observations and confirm whether the event recurs;

- Conduct a thorough literature review and read everything known about the topic;
- Ensure you will be able to carry out a robust study with this animal, under the field conditions in which it lives and during its active periods, and that you will not face personal limitations in pursuing this research;
- Discuss with colleagues and make sure you will have field companions.

This initial stage is fundamental to your success. Imagine if Raoni had not had field companions to work with in the heart of the Amazon rainforest? It can be even more dangerous to work on floral biology and pollinators in a large city such as São Paulo, for example. Thus, knowing the animal, ensuring the topic has not yet been studied, having field companions, and ensuring safety are the basic requirements for starting your study.

The next step is to define the methodology:

- How will I observe this animal?
- What equipment should I use?
- What should the observation sessions be like?
- How long should I observe for?
- How should I record my observations?
- Do I need to mark the animals?

This stage of defining the methodology is crucial. A mistake here and everything will be lost; unreliable data are of no use. To determine which methodology to use, once again you will need to consult the literature. Therefore, a thorough literature review at the initial stage is important. It is also important to talk to more experienced colleagues. Imagine what it would be like for someone who has never worked in the Amazon to conduct a study on small fish in forest streams without support, help, or advice. It would be impossible. Build friendships in the scientific and academic environment, avoid problems, distance yourself from people who do not contribute, who show no altruism or empathy. Academic success comes with healthy partnerships, constructive criticism, humility, and perseverance.

Very well. You already have a question, you know it is interesting, and you are defining a methodology based on the literature and your research question. This question, or questions, should also be supported by a hypothesis. Simple questions, with direct and clear hypotheses, are much more likely to yield good results. Simplicity is usually accompanied by sound reasoning and good decision-making. In science, we use the classic hypothetico-deductive method, which involves formulating specific hypotheses based on observations or available information, followed by the logical deduction of consequences that can be empirically tested. Based on these tests, the hypotheses are confirmed, modified, or rejected.

The classic definition of the *hypothetico-deductive method* is attributed to Karl Popper in his work *The Logic of Scientific Discovery*, published in 1934 [3]. Popper described the hypothetico-deductive method as a process in which scientists formulate hypotheses that are then subjected to rigorous and critical testing. Scientists emphasized the importance of falsifiability, arguing that a scientific theory must be testable and potentially refutable by empirical data. Thus, in the context of hypothesis testing, the null hypothesis (H_0) is an assertion that is initially assumed to be true. From this null hypothesis, an alternative hypothesis (H_1) is formulated, representing an opposing situation or a significant relationship between variables. For example, in the case of the small fish, we assume as H_0 that the different colourations of *A. hyppolitae* **are not** associated with different types of behaviour. Our alternative hypothesis (H_1) is that the different colourations of *A. hyppolitae* **are** associated with different types of behaviour.

However, in Animal Behaviour and Behavioural Ecology, we do not always work with simple paired hypothesis tests. Often, the behavioural nuances are such that we must resort to the *multiple hypothesis testing* approach. Multiple hypothesis testing, also known as multiple comparisons testing, is a statistical approach used when one wishes to simultaneously compare several null hypotheses with their respective alternative hypotheses. Let me give you an example.

6.2 Multiple Hypothesis Testing Method

In the previous chapter, we briefly discussed the study by [4] on the wasp *Brachygastra lecheguana*, which specializes in preying on endophytic beetle larvae that develop inside flower buds. In this study, we considered hypotheses that these wasps used: (a) visual cues (beetle holes in the buds); (b) chemical cues (beetle scent on the buds); or (c) a combination of different mechanisms, such as visual, chemical (odour), and possibly tactile cues, to locate the beetle larvae.

For the study, it was necessary to use a combination of experimental manipulations (visual, chemical, and visual/chemical) directly in the field (Fig. 6.2). Note the importance of clearly defining what will constitute the treatment and control groups in our experiments. An experiment without a control is a flawed experiment. In our example, if I am going to pierce the bud, my control should be to touch it with a pin without piercing, so that a critic cannot claim that the mere touch had an influence. Or, if I am going to soil the bud with larval macerate (treatment), I should have a control using only water (control), so that the bud undergoes the same stress of being touched and smeared or soiled with a brush. Think of the journal reviewer as the "devil's advocate."

Our results supported the hypothesis of [5] that wasps, including the *B. lecheguana*, employ a combination of visual, chemical, and possibly tactile cues to locate endophytic beetles within floral buds. We suggest that the wasp uses a sequence of vision, olfaction, and touch to find the concealed larva. We observed that wasps landed significantly more often in the experimental treatments (visual, chemical, and visual/chemical) than in the control treatments, indicating that our hypothesis "c" was supported, i.e., wasps use "*a combination of different mechanisms, such as visual, chemical (odour), and possibly tactile cues, to locate beetle larvae.*"

A well-designed study must present clear and well-defined hypotheses, premises, and research questions. If you do not see the relationship between your hypotheses, questions, and some aspect of the scientific theory directly related to your study, something is amiss. Either you are not studying what you should, or you are not doing it

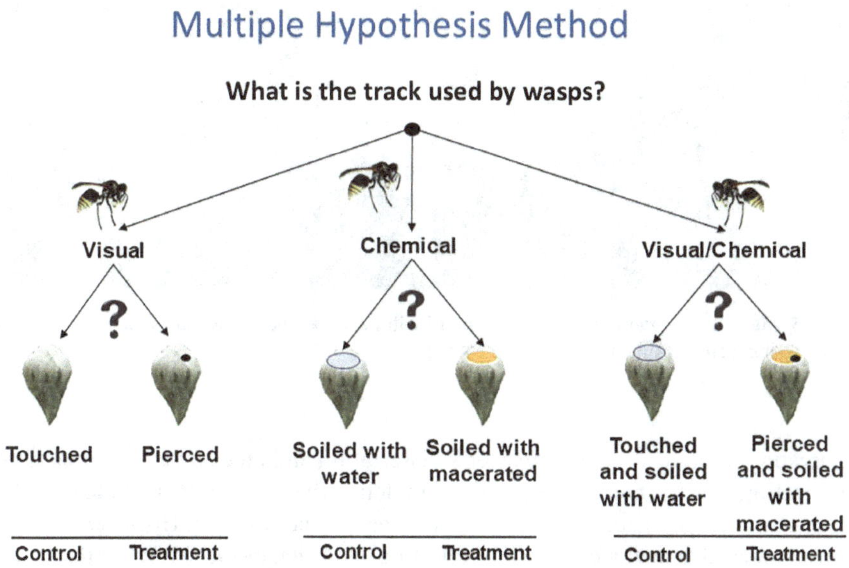

Fig. 6.2 Multiple hypotheses method in Animal Behaviour. Experimental manipulations in the field to test multiple hypotheses regarding the mechanisms used by the wasp *Brachygastra lecheguana* to locate and prey on endophytic Curculionidae beetle larvae in Malpighiaceae floral buds in the Cerrado. For this, buds were touched with a pin (control) or pierced (to simulate beetle holes—treatment), soiled with water (control) or with a macerate of beetle larvae to simulate odour (treatment), or a combination of visual and chemical manipulations was applied. Adapted from [4]

properly, or you may be poorly supervised. On the website *The Science We Do* (blog), you will find articles on the importance of good supervision. Take a look! It may help.

You now understand that you must have a straightforward research question, robust hypotheses grounded in the literature, and well-defined methods. Let us move on to the fieldwork and data collection. Once again, tormenting questions arise! One of the main ones: what should my sample size (N) be? That is, how much data should I collect for each variable, how many individuals should I mark, for how many hours, or in how many sessions should I observe? How do I know when it is sufficient, when it is correct? These doubts are daunting and keep young researchers awake at night, as they did my Peruvian student when working with dragonflies [6] (Fig. 6.3).

In male odonates, both body size and fat content are related to territorial defence and reproductive success (see Chapter 3). Larger males with greater energy reserves are more successful in territorial disputes and attract more females. Wing colour has also been considered a mechanism influencing agonistic behaviour among males, as wing pigmentation may serve as a signal of male quality. In this study, I proposed to the student that we analyse the hypothesis that "a set of physical (body size and wing colour), physiological (body fat content), and behavioural (male–male contests) characteristics of the neotropical dragonfly *Zenithoptera lanei* (Anisoptera: Libellulidae) determine their success in territory defence and mating behaviour." This highly

Fig. 6.3 Male *Zenithoptera lanei* (Anisoptera: Libellulidae) perched on a plant stem within its territory, opening and closing its beautiful blue wings

interesting topic resulted in a master's dissertation and an excellent international publication, but not before we struggled to define the number of individuals to be used, observed, marked, and tested at each stage of the research. However, it is not only numbers that matter during data collection. Note that the fieldwork for this study was conducted from November to January, between 9 am and 2 pm. Why? You must collect your data at the times and during the period when your study object (animal or subject) is available in the field or laboratory.

And who helps you with this **are three great allies**:

- the **literature**—bibliographic references, colleagues who have already worked on this topic;
- **preliminary and confirmatory field observations**;
- **colleagues and/or your supervisor**.

In one stage of this study, we measured the body fat content in 80 males (29 territorial, 29 subordinate, and 22 males collected after mating behaviour). To define these numbers, we had to rely not only on the literature, but also on our collaborator colleagues with the greatest expertise in statistics, to determine which analytical methods we would employ and what the basic requirements were, indicating minimum sample sizes. Indeed, during the methodology phase and at the start of data collection, it is essential to define which statistical tests will be used and to be aware of their limitations, their basic assumptions, and the conditions required for statistically robust analysis. The success of research in our field depends on extensive study, reading, and, above all, collaboration. It was this combination of factors that enabled us, in this study, to demonstrate that males of *Z. lanei* could be characterised as territorial when they repelled other males and remained in the same location within the lake for two consecutive days. In general, these territorial males were larger and had more abdominal and thoracic fat, engaged in chases more frequently, spent more time in sexual behaviour and female guarding, and mated more often compared to subordinate males. When assessing whether the percentage of wing area covered by black paint (one of our experimental manipulation tests to investigate the function of wing colouration) influenced male behaviour, we found that territorial males tended to act aggressively towards other males whose wings were partially painted, and sexually towards females, regardless of the painted wing area. In *Z. lanei*, both body size and fat content play a role in determining territoriality. By dominating competitors and occupying preferred sites in high-quality areas, these males are more likely to be visited by females and to engage in mating.

A point of utmost importance during the data collection phase is the organisation of these data. Between the data collection phase and the subsequent phase, which is the analysis of the collected data, there is a continuous and intermediate stage that often remains overlooked, as if it were unnecessary, but which may in fact be one of the most important in any study: ***data organisation***. This is necessary for several reasons, but primarily because:

- well-organised, well-described, and promptly shared data—whether stored in the cloud or as a physical electronic file with the research team—become secure data. When they are well described, safely stored, and backed up, you can be confident that you or someone else will complete the study, which may contribute to the knowledge and conservation of a species;
- data organisation allows for their early arrangement in spreadsheets, saving time in future analyses and enabling a more accurate determination of whether the minimum sample size required by the statistical tests to be used has been achieved;
- behavioural observation data should be recorded immediately, reviewed and corrected at the time of observation, properly written up, and stored. One cannot rely on memory and leave it for later, as the likelihood of errors that could compromise the work is high.

In Animal Behaviour, if you have not carefully observed and studied all the steps described so far, you may make many mistakes that could compromise your research and your scientific reputation (Fig. 6.4).

A behavioural observation must be repeated several times to ensure that the observer is truly seeing what is happening. The observer must be mindful of their clothing, noises, scents, and anything else that might stimulate or inhibit the behaviour of the animal under observation. As mentioned in previous chapters, the sensory world of animals is very different from our own, and we must always take this into account. Imagine entering the forest wearing a red shirt in the hope of observing birds or mammals. I would be seen from a great distance and the animals would flee or hide. This would certainly lead to observation errors (**type I error**). It is more common than one might think for a behavioural act not to be exactly what we assumed, or, at times, to be something entirely different from what we first imagined. See Fig. 6.5 and, before continuing with the text, try to guess what is happening.

Anyone looking at this photo (Fig. 6.5) would imagine that these tiny "scorpions" are going to kill and eat this bug. And that is indeed what will happen. So, well done—you were correct in your observation!

But… If the observer is a little more patient and watches more carefully, as we did [7], they will see the hemipteran take flight with the pseudoscorpions attached to it. Following its flight, we saw the insect land on another tree several dozen metres away. On this other tree, the animal weakens and ends up being subjugated by predators,

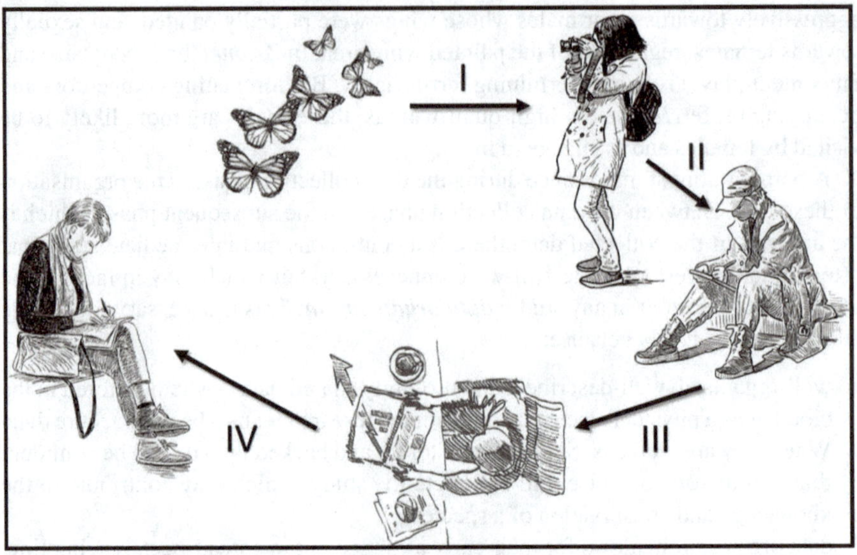

Fig. 6.4 The four interconnected types of error most commonly made in animal behaviour studies: I—observation error; II—recording error; III—analysis error; and IV—interpretation error (free images: www.depositphotos.com)

Fig. 6.5 Group of pseudoscorpions, *Paratemnoides nidificator*, attacking the legs of a hemipteran. Photo by Everton Tizo-Pedroso

which now occupy a new tree, far from the original colony. For such a small animal with limited dispersal abilities, this journey is akin to crossing a continent. This behaviour of using other animals for transportation, like hitchhiking, is known as phoresy. Phoresy (attachment to a carrier animal) is the primary mode of dispersal in *P. nidificator*. The clustered individuals use large animals as dispersal vectors and, once established, use the vector as the first food item for the new colony. Thus, in Fig. 6.5, we are witnessing the beginning of a phoretic process, which will culminate in predation and the establishment of a new pseudoscorpion colony in a tree far from its original colony.

An observation error (Type I error) will lead to a **Type II** error, a recording error. Poorly made notes, a rushed transcription from a recording to a table, lack of attention, excessive joking and distractions in the field or laboratory, are other common factors that lead to recording errors. Diligence is the remedy here! Those who work with care, dedication, and know when to work and when to play, rarely make gross mistakes. Once again, having a good understanding of the animal, its capabilities, as well as its environment—knowledge acquired through reading, preliminary observations, and discussions with more experienced colleagues—is fundamental.

Clearly, an observation error leads to a recording error, which in turn leads to an error during data analysis, or a **Type III** error, an analysis error. You may have done everything correctly up to this point and still make a mistake in the analysis, simply by not choosing the appropriate statistical test, by being unaware of the limitations of the test you are using, by not testing for data normality, or by not considering the type of mathematical distribution involved. Many factors can interfere at this stage, during analysis. I strongly recommend reading, examining how others have handled similar data. I suggest consulting a statistician familiar with biological analyses, or more experienced and patient colleagues. Yes, patient, because if they are to teach someone who struggles with statistics as much as I do, they must be patient. But a colleague may have done everything absolutely correctly up to this point and, due to lack of knowledge—that is, due to insufficient reading, lack of understanding of the study animal, related species, or the subject—make a misjudgement, that is, commit a **Type IV** error, an interpretation error.

Misinterpreting a set of analyzed data is disastrous; it can compromise a dissertation, a thesis, or a publication, and may tarnish a career depending on the extent and reach of the error before it is detected. That is why I strongly recommend that you do not become a mere collector of PDFs, but that you read! Study and discuss with colleagues, exchange information about your research topic.

One aspect that greatly concerns me in science is these errors, which are always worrying. They may occur by accident, by chance, or by misfortune, but we must not allow them to happen through negligence, carelessness, lack of attention, diligence, or dedication. Rigour in science is essential. Often, it is better to redo an entire study, or even abandon it if we are not certain whether there was any error or omission, than to persist and publish something in which we do not have complete confidence. Preserving our reputation is fundamental. Science is one of the few activities in the modern world where a name still holds value, simply for being a name—a name untainted by any failure, fault, unethical or immoral act. In other words, your name in science always has value, until proven otherwise. Preserve your name.

Unfortunately, we have seen both young and senior researchers have their articles retracted and removed from journals after publication. This is very bad, very sad, and extremely damaging to science as a whole. Therefore, take care of your name, your career, and those with whom you associate. Recently, a case became emblematic of bad scientific conduct: a young and prominent researcher was caught falsifying data in his studies. Two dozen of his publications were reviewed and retracted, removed from the journals in which they had appeared. I can only imagine his co-authors, who knew nothing of the falsifications and whose names were dragged into oblivion along with his. Think about this: lies have short legs, and in science this is truer than ever.

Once the data are organized and reviewed, with no errors and with every detail attended to, we move on to the analyses! Then comes that little word that causes panic for many: statistics!

Statistical analysis plays a crucial role in the study of animal behaviour. It guides the process from data collection to analysis, providing an objective and evidence-based understanding of animal behaviour. There are several ways in which statistical analysis is important in the study of animal behaviour:

(a) it allows for the correct description and summary of the data collected in a precise and concise manner. This includes calculating means, standard deviations, frequencies, and other statistical parameters that provide quantitative information about animal behaviour;
(b) the identification of patterns and trends. For example, with statistics, we can determine whether certain behaviours occur more frequently at specific times of day or whether there are significant differences between groups of animals regarding certain behaviours;
(c) hypothesis testing: statistical analysis allows the testing of specific hypotheses regarding animal behaviour, thereby validating or refuting assumptions and theories about it;
(d) it can also be used to construct mathematical models that describe animal behaviour. These models can be employed to predict future behaviour based on historical data, providing a valuable tool for understanding and anticipating animal behaviour in different situations;
(e) it enables the comparison of groups of animals or different experimental conditions. This is useful for determining whether there are significant differences between groups and whether certain factors consistently influence animal behaviour.

In summary, statistical analysis plays a crucial role in Ethology, as it provides a solid foundation for the scientific understanding of animal behaviour.

Studying statistics, its foundations, its applications, and its tools is as important as mastering the English language (If you are a native English speaker, imagine the difficulty for those who are not—empathy is a gift in science).

The good news is that statistics is not the monster it is often made out to be. Frequently, our aversion to statistics or mathematics stems from a lack of dedicated study, or from having encountered a teacher lacking in didactic skills or patience. If you wish to advance in science, you will need to learn at least the basics of statistics and collaborate with those who are proficient in this highly relevant field.

There are several **common statistical analysis errors** that can lead to misinterpretation of results. Therefore, to ensure your study is reliable, you must be alert to these **errors**. The **three main** ones are:

(1) failing to consider sample size (insufficient sample size)—small samples may yield results that are not representative of the population under study, which can lead to misleading conclusions and incorrect generalisations;
(2) inappropriate use of statistical tests—each statistical test has specific assumptions and conditions that must be met for its correct application. Using the wrong statistical test for the research question or ignoring the necessary assumptions can result in incorrect outcomes;
(3) confusing correlation with causation—it is a common mistake to assume that a correlation between two variables implies a direct causal relationship. Correlation merely indicates a statistical association between variables, but does not establish a cause-and-effect relationship. Care must be taken when interpreting relationships found in the data, and causal claims should be avoided without additional evidence or controlled experiments. One should always be mindful of the true biological significance of a result.

Currently, the most popular programme for statistical analyses in biology is the R Core Team. (2021). R: A Language and Environment for Statistical Computing. R Foundation for Statistical Computing, Vienna, Austria. URL: https://www.R-project.org/. However, within a few months or within a year, I believe that something entirely new and more user-friendly will emerge, based on what we are seeing with Artificial Intelligence (AI) programmes. An example is ChatGPT, launched at the end of 2022: OpenAI. (2021). ChatGPT [AI Model]. https://openai.com/

Once your data have been properly analysed and checked, including statistical figures, illustrations, and videos, it is time to interpret all the results and knowledge you have accumulated. For this, you must be thoroughly up to date with the literature, both classical and recent. The literature, supported by discussions with more experienced colleagues or peers, will greatly assist in understanding your findings and their true biological significance.

Do not hesitate or feel embarrassed to seek help or to write to more experienced colleagues requesting support in discussing the data you have analysed. I have done and continue to do this frequently. It is common for me to write to colleagues, both nationally and internationally, with whom I have had no prior relationship, but whom I know are prominent in my field, seeking support. In general, the response and expertise are excellent, with rare exceptions. I have many international collaborations as a result of this approach—seeking help, discussing, presenting new ideas. This has opened many significant doors for my students. Today, I have former students in the United States, Spain, Canada, Australia, Finland, England, Mexico, Peru and Israel, who are highly successful, having taken their first steps from a simple email, a cordial message presenting a problem of mutual interest and proposing collaboration. In this regard, the Coordination for the Improvement of Higher Education Personnel (Capes) and the National Council for Scientific and Technological Development (CNPq) have been fundamental, providing scholarships for my students and funding research projects that enabled some of our trips abroad to establish partnerships.

Constructive criticism at all stages of your planning should be welcomed and valued. Especially in the interpretation of analysed results, such criticism can be crucial for a successful publication, as it may anticipate questions that would arise from a reviewer. In summary, to ensure correct interpretation of data and to follow a rigorous approach based on scientific principles, we must understand the context; that is, before beginning to interpret the data, it is essential to understand the context in which they were collected. This includes knowing the methodology used for data collection, the limitations of the study, and the specific research area. It is essential to gain a detailed understanding of the data collection procedures, the statistical analyses performed, and the results obtained. Make sure you understand the statistical methods used. All studies have limitations, and it is important to identify them. Consider potential biases, methodological issues, or limitations in the interpretation of the results. This will help you develop a more critical perspective on the data. Understand that scientific data must be interpreted within a broader context. Consider other studies related to the same topic or model organism and assess whether the results are consistent with other available scientific evidence. Discuss with your peers. Accept criticism. A well-designed research project follows a long and rigorous path (Fig. 6.6).

Generally, a research project concludes with a series of new questions. Learn to control your urge to answer yet another question. Strive to be objective before continuing with the same topic; in other words, publish your results. You may uncover the love life of Amazonian dwarf cichlids, but first you will need to answer how they

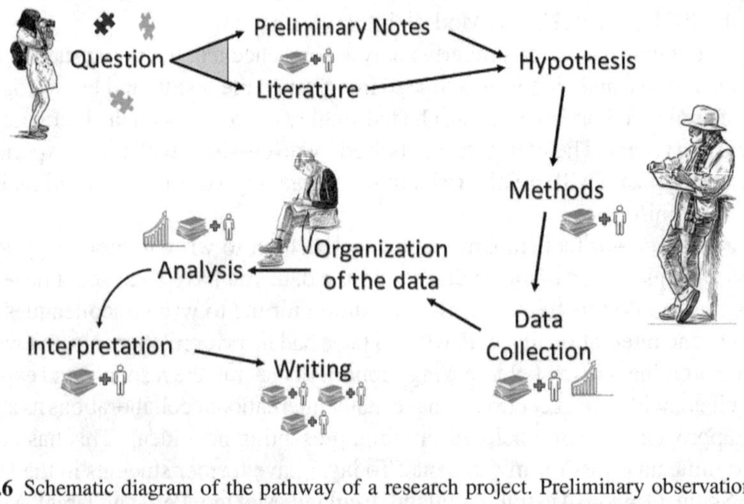

Fig. 6.6 Schematic diagram of the pathway of a research project. Preliminary observations, followed by extensive reading and discussion with peers (books + people), lead to sound and valid hypotheses. A well-defined methodology, followed by rigorous data collection underpinned by statistical analysis (graph), leads to correct analysis, interpretation, and writing

swim, what they eat, where they reproduce, and each of these questions can become an important published work. Brick by brick, you build the foundations of your research line and your career.

Chapter Challenge
Choose a published scientific article in the field of Animal Behaviour that interests you. Read the article. Then, try to work backwards:

- How might the project that led to this article have been designed?
- What were the foundational texts, the central questions?
- What hypotheses and related theories gave rise to this study?

This is an outstanding exercise. You will gain a great deal from it.

Challenge Objective
Knowing how to design a research project is an essential part of our career. Without a project, there is no research. Without thinking about hypotheses and how to test them, there is no good project. Practise, practise, and practise—this is the way forward. Dedicating yourself to this exercise will greatly enhance your ability to develop in science.

Go Further
I have prepared a video about the study with the wasps presented in this chapter: https://youtu.be/0GOpnbSoA-4?si=4M5-kLVhRtNBHf8u.

References

1. Rodrigues RR, Carvalho LN et al (2009) Colour changing and behavioural context in the Amazonian Dwarf Cichlid *Apistogramma hippolytae*. Neotrop Ichthyol 7:641–646
2. Rodrigues RR, Zuanon J et al (2012) Reproductive behaviour of the Amazonian dwarf cichlid *Apistogramma hippolytae* Kullander, 1982: offsetting costs and benefits. Acta Ethol 15:47–53. https://doi.org/10.1007/s10211-011-0107-8
3. Popper K (1934) The logic of scientific discovery. Routledge, London
4. Mendes-Silva I, Queiroga D et al (2021) Multiple cues guarantee successful predation by a Neotropical wasp. Behaviour 159(7):643–655. https://doi.org/10.1163/1568539X-bja10144
5. Torezan-Silingardi HM (2011) Predatory behaviour of *Pachodynerus brevithorax* (Hymenoptera: Vespidae, Eumeninae) on endophytic herbivore beetles in the Brazilian tropical savanna. Sociobiology 57(1):181–185
6. Valdivia FGA, Alves-Silva E, Del-Claro K (2020) Differences in size and energy content affect the territorial status and mating success of a neotropical dragonfly. Austral Ecology 45:748–758
7. Tizo-Pedroso E, Del-Claro K (2007) Cooperation in the neotropical pseudoscorpion, *Paratemnoides nidificator* (Balzan, 1888): feeding and dispersal behaviour. Insect. Soc. 54:124–131. https://doi.org/10.1007/s00040-007-0931-z

Chapter 7
The Functional Question ("How?") and the Evolutionary Question ("Why?")

This chapter will feature many stories. Understanding the history and the processes involved in a successfully completed study is, in my view, an effective learning tool.

To understand the leap from basic studies of Animal Behaviour to modern Behavioural Ecology, with the use of experimental manipulation, phylogenetic comparison, genetic tools, and sophisticated measurement and analysis equipment, examples are extremely helpful. They will be particularly useful in this chapter, in which, after presenting the tools, we will learn how to use them.

7.1 Hypothesis Testing in Behavioural Ecology: The Ants and the Nectar

Relationships between ants and plants have made significant contributions to our understanding of the functioning and conservation of natural communities. Ants can have various interactions with plants, and, in most cases, these relationships result in some benefit for the plant, the ant, or both. The most common resource that plants offer to attract ants is extrafloral nectar (EFN), a liquid substance rich in carbohydrates with diluted concentrations of amino acids, lipids, phenols, alkaloids, and volatile organic compounds. Carbohydrates have been suggested as key resources for arboreal ants. Extrafloral nectaries (EFNs) are found in at least 93 plant families across 332 genera worldwide, and are very common in our Cerrado (Fig. 7.1).

The earliest study discussing the function of EFNs dates back to the late nineteenth century. The study, entitled *The relations of ants to the plant world*, was conducted by the American entomologist William Morton Wheeler and published in 1891. Although the study focuses mainly on interactions between ants and plants, it also discusses the presence and function of EFNs. Wheeler [1] observed the presence of EFNs in a wide variety of plant species and investigated the behavioural responses of ants to these

K. Del Claro, *Behavioral Ecology*, https://doi.org/10.1007/978-3-032-13988-7_7

Fig. 7.1 The ant *Ectatomma tuberculatum* with mandibles full of nectar collected from extrafloral nectaries at the base of *Qualea multiflora* (Vochysiaceae) leaves in the Cerrado. Photo: Denise Lange

nectar sources. He described the morphological characteristics and specific location of the nectaries, as well as their effects on the behaviour of ants and herbivores. As can be seen, Wheeler provided a detailed account of "**how**" this ant behaviour unfolds, that is, the functional aspects of this behaviour: how the ant climbs the plant, whether it drinks the nectar, and whether it interferes with herbivores and other animals on the plants.

This pioneering work by Wheeler laid the foundation for subsequent studies on extrafloral nectaries and their ecological importance. Since then, numerous researchers have explored the various functions of EFNs, but with an almost exclusive emphasis on the effect of ant presence on plants, with the vast majority of these studies confirming such benefits (e.g., [2]). Thus, for over a century, the benefits of this association for ants have been taken as established fact. However, no specific Behavioural Ecology studies have tested the adaptive value of ant feeding on EFNs for individuals and colonies. In other words, "**why**" do ants feed on EFNs? What advantages of this behaviour for ants justify its evolutionary selection? Noting this gap in the ecology of interactions, we decided to conduct a behavioural study to determine whether feeding on EFNs is indeed beneficial for ants. We hypothesised that "*ants feed on EFNs* ***because*** *this significantly increases their adaptive value,*" for example, by accumulating more fat, increasing colony size, and improving individual survival.

A simple, clear question that seems easy to answer, doesn't it? So why has no one done this in a century of research? Why was a hypothesis simply accepted as fact? Was it because it was difficult to test in practice? We will show you how we did it and let you decide.

We chose the arboreal ant *Cephalotes pusillus* (Myrmicinae) as our model, as it is very common in our study area and neither bites nor stings. Moreover, it is a frequent visitor to EFNs of trees and shrubs in the Cerrado, such as *Chamaecrista desvauxii* (Caesalpiniaceae) (Fig. 7.2). This plant species is easy to handle, and its seeds

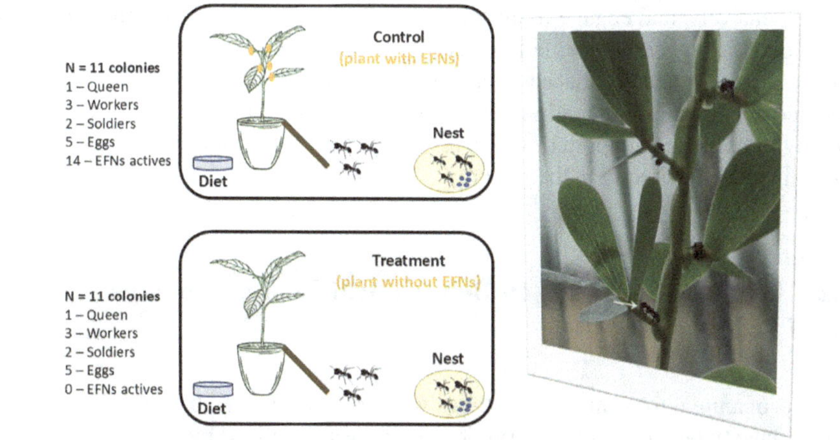

Fig. 7.2 Experimental design to test the hypothesis that *Cephalotes pusillus* ants that feed on the EFNs of *Chamaecrista desvauxii* significantly increase their fitness. We hypothesised that ants feeding on nectar would be heavier and their colonies more populous

germinate quickly. In summary, we collected seeds in the field and raised seedlings of this plant in the laboratory. After 45 days, with the seedlings already established and with active EFNs, we returned to the field to collect ants. Ant nests were collected and kept in the laboratory in 22 boxes, each containing a nest with one queen, three workers, two soldier ants and five eggs (Fig. 7.2). These 22 boxes were then randomly assigned to 11 boxes for the "control" group and 11 for the "treatment" group. In both groups, the ants were provided daily with a small container containing 5 g of a specific diet for ant rearing (water, proteins, carbohydrates, etc.). In addition, in each box or experimental arena, a seedling of the plant *C. desvauxii* was introduced. The difference was that, in the control group, all seedlings had 14 active EFNs, whereas in the treatment group, we removed all EFNs from the plants (details in [3]). Once the experiments were set up, over the course of a year, we counted all individuals, queens, workers, soldiers, eggs, and finally weighed the colony and its individuals.

A good study in Behavioural Ecology, such as the one in this example, must be scientifically grounded in theories and hypotheses that the author seeks to corroborate or refute. The mere description of behaviours is a useful tool in more complex studies, but on its own, it is currently insufficient to warrant acceptance for publication in a reputable indexed journal. The indexing of scientific journals is a process by which academic periodicals are evaluated and included in specialised databases or indices. The most well-known and widely used databases and indices for scientific journal indexing include Web of Science, Scopus, PubMed, Medline, among others. For a journal to be indexed in a database, it generally needs to meet certain quality criteria, such as peer review, scientific rigour, regular publication, compliance with ethical standards, and other specific requirements set by each database. Therefore, to achieve a good publication, you should follow the steps outlined here: plan carefully, read extensively, and engage in thorough discussion and reflection with your peers, collaborators, and colleagues before getting started. In the example with the ants and EFNs, the results showed that a nectar-rich diet (providing at least 30 calories per day) resulted in five times more individuals per colony, greater body mass, and more eggs. Our hypothesis was supported. If you are interested in this topic, we later tested the same hypothesis directly in the field [4].

7.2 Hypothesis Testing in Behavioural Ecology: Beetles and Lizards

Aspects of animal form and colouration always attract considerable attention, such as camouflage (when an animal is cryptic, i.e., it goes unnoticed by a visually oriented predator—Fig. 7.3A); aposematism through warning colouration (when an animal signals unpalatability, toxicity, or danger through its colours, sounds, behaviours—Fig. 7.3B); and mimicry, when a palatable animal imitates an aposematic model (Fig. 7.3C, D).

Fig. 7.3 Butterfly camouflaged as a dry leaf in the Cerrado (**A**), aposematic frog (*Brachycephalus*) (**B**), and model wasp (a toxic or unpalatable aposematic animal), *Entypus ferriginipennis* (**C**), which is mimicked by the palatable bush-cricket, *Scaphura nigra* (**D**)

My first published study [5] on a case of mimicry involved an Aticinae beetle, *Homophoeta octgoguttata* (flea beetle), and its mimic, a Cerambycidae beetle, *Adesmus colligatus* (longhorn beetle). I will use this study as a second example.

Observing these animals in the field (Fig. 7.4), I noted that they were very similar in morphology (oval and small, 4 mm in length) and colouration (yellowish-green, with black chequered patterns on the elytra), occurred in the same habitats (understorey shrubs, trails, and clearings), at the same times of day (from 8 am to 3 pm), and during the same season (summer). The alticine beetle had the third pair of legs modified for jumping, which made it more difficult for visually oriented predators to capture. Studies by other colleagues indicated that these flea beetles were unpalatable, i.e., they contained toxic substances in their bodies (alkaloids). Therefore, they were good models for other insects wishing to mimic them, taking advantage of their resemblance to a noxious animal to deceive shared predators.

In a preliminary field observation, I hypothesised that this was a case of Batesian mimicry, like that observed between wasps and bush-crickets (Fig. 7.3). This type of mimetic relationship was first proposed by the English naturalist Henry Walter Bates in 1862, after ten years studying butterflies in the Amazon. In this relationship, a "noxious" animal, that is, one possessing some unpleasant characteristic for its predators—such as the presence of toxins, spines, a sting, or any feature that hinders its capture, ingestion, or swallowing—signals this to its predators through conspicuous colouration, behaviour, or another obvious trait. This "unpalatable" animal serves as a "model" that other animals, which are not in fact noxious or unpleasant to their predators, imitate. The imitator, which resembles the model in colouration, morphology, and behaviour, is known as the "mimic." In fact, this is a case of parasitism, as the mimic stands only to gain while the model stands to lose, since confusion on the part of predators may result in the death of a model individual.

A simple visual comparison between the two beetles, the model (Alticinae) and the potential mimic (Cerambycidae) (Fig. 7.4), analysing colour, shape, and size, indicated their similarity. The overlap in habitat, activity period, and potential predators further suggested that this could indeed be a case of mimicry. However, in science it is not

Fig. 7.4 Beetle (Coleoptera), Chrysomelidae, Alticinae, *Homophoeta octgoguttata* (**A**) and Cerambycidae beetle, *Adesmus colligatus* (**B**)

enough to state what we imagine. **We must prove our ideas**, and this is where hypotheses and predictions (the hypothetico-deductive method) come into play, guiding the experimental methodologies of Behavioural Ecology. The **hypothesis** proposed in this experiment was that "***Adesmus colligathus*** **is a mimic of *Homophoeta octoguttata*, whose mimetic resemblance to the model increases its chances of survival against visually oriented predators.**"

To test this hypothesis, I had to consider the predictions, also called premises, which would strengthen the supposition if confirmed, or weaken it if rejected. A **first premise** considered that "**there would be a visually oriented beetle predator in the study area, frequenting the same environments as mimics and models at the same times.**"

Field observations indicated that there were dozens of bird species that could be considered potential predators, but most were difficult to capture and handle in captivity. In the field, I also noted that lizards could be important beetle predators in the area where these insects occur. From the literature and discussions with colleagues, I discovered that Tropiduridae lizards, such as *Tropidurus itambere*, were common in the study area and commonly fed on insects such as beetles and ants. They were easy to capture and handle in the laboratory, an ideal condition for conducting a series of experiments. Thus, the first step was accomplished: I had a known and captured predator. The second step would be to demonstrate the unpalatability of the model, that is, its unpleasant taste. Thus arose the **second prediction**: "**Alticinae models are unpalatable.**"

Several *H. octoguttata* were collected and offered to seven-day-old chicks (*Gallus gallus*) (inexperienced predators) and to the lizards. The beetles were rejected, and subsequently the predators were offered larvae of beetles known to be palatable (*Tenebrio molitor*). These were eaten immediately, confirming the unpalatability of the alticine beetle. With this stage completed, we moved on to a **third premise**: "**The mimic beetles are palatable and, therefore, can be preyed upon by the lizards.**"

Some of the longhorn beetles, supposedly mimetic, were offered to chicks and lizards that had had no prior contact with the model. The longhorn beetles were promptly attacked and eaten, confirming their palatability.

At this stage of the research, three important steps have been clarified for confirming the mimicry hypothesis:

a. the models (Alticinae) are unpalatable;
b. the mimics (Cerambycidae) are palatable;
c. in the same area of occurrence, there are visually oriented beetle predators that recognise the unpalatability of the models.

The next step was a key manipulation experiment to confirm or refute the **fourth premise**: "**Visually oriented predators confuse mimics with models and reject them.**" To this end, I devised the following experimental sequence (Experimental Manipulation, Fig. 7.5):

1. the lizards are given a palatable beetle larva (*T. molitor*)—interval (5 min);
2. the same lizards are presented with a model beetle—interval (5 min);

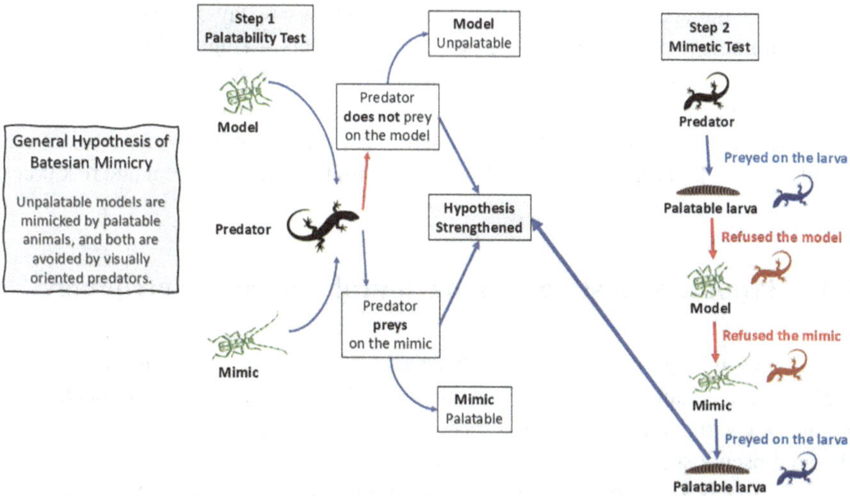

Fig. 7.5 Example of an experimental manipulation test involving a case of Batesian mimicry. The first step is to test the unpalatability of the model to a predator, followed by testing the palatability of the mimic (Step 1). An unpalatable model should be rejected by a naïve predator, while a Batesian mimic should be consumed. In a mimicry efficacy test (Step 2), a palatable animal is offered to the predator; it should be consumed. Next, a Batesian model (an unpalatable animal) is offered, which should be rejected. After this test, the same predator should be given a Batesian mimic and also reject it, thereby demonstrating visual confusion with the "bad" model. Finally, a palatable animal is offered, which should be consumed by the predator, showing that it was hungry but actively avoided both the model and the mimic. See more in [5]

3. the same lizards are now presented with a mimic beetle—interval (5 min);
4. Finally, the same lizards are offered a palatable beetle larva (*T. molitor*).

To support the mimicry hypothesis, the following predictions had to be met:

1. the lizards had to eat the palatable larva both before and after exposure to models and mimics. This would demonstrate that they were neither hungry nor already satiated, respectively;
2. the lizards had to reject both the model and the mimic. This would demonstrate that a "choice" was made.

And this is exactly what occurred. The naïve lizards, when attempting to capture the models (Alticinae), spat them out, repeatedly wiped their mouths on the sand at the bottom of the terrarium, and did not approach the prey again. When the *Tropidurus* encountered the mimics (Cerambycidae), they approached, examined, and then withdrew, thus rejecting the mimics. Visually oriented predators were indeed confused by the mimics, mistaking them for the models. To clarify this example of experimental manipulation using a classic Batesian mimicry test, refer to the proposed diagram (Fig. 7.5). It illustrates the basic sequence of a palatability test, followed by a mimicry test with a positive result.

Note that in the examples presented, I show you how a behaviour occurs, including its morphological, physiological, and behavioural aspects—a highly descriptive, functional phase. In the second part, we test the adaptive value of the behaviour, whether it increases the survival or reproductive success of the animal displaying it, thereby influencing future populations and the evolutionary aspects of behaviour. It is entirely possible to work in Behavioural Ecology, even with limited equipment and resources.

7.3 Hypothesis Testing in Behavioural Ecology: Matriphagy

Often, those who go into the field or observe an animal or group of animals exhaustively have the chance to witness rare, unexpected events that can reveal much about the biology and natural history of a species. This was the case with our observations of social pseudoscorpions.

There is a Neotropical species, previously mentioned in this book (Chapter 5), *Paratemnoides nidificator* (Atemnidae), which lives in colonies beneath tree bark in forests and the Cerrado. Colonies are composed of adults (reproductive and non-reproductive males and females) and nymphs (across four stages), as well as eggs. Females weave reproductive chambers with silk for egg incubation and juvenile care. Non-reproductive females, and especially nymphs, weave moulting chambers. These animals live together, hunt cooperatively, display parental care, and have a complex social life [6].

In laboratory-maintained colonies, we observed that females wove the reproductive nest alone. Inside the chambers, they provided continuous parental care to embryos and nymphs, feeding and cleaning them. This continued until the nymphs became adults or were expelled from the nest. Females also guarded the nest entrance against intruders. In some cases, they produced an additional brood and then forced the nymphs from the first brood out of the chamber by touching them with their pedipalps. After driving out the original brood, the females sealed the exit with fresh silk and produced a new brood sac. The "displaced" nymphs then cooperatively constructed another chamber in which they moulted. The reproductive and moulting chambers were built side by side, sharing vertical walls. Other important female behaviours included personal hygiene and feeding. Males directly cooperated in parental care by capturing and offering prey to all colony members. However, in general, nymphs were fed by the mother. Our data revealed that females left the nest to hunt before feeding the brood. Mothers ate the remains of prey left by the nymphs. Tritonymphs (pre-adult nymphs) hunted cooperatively with or without the mother [7].

However, in our observations, in addition to these fascinating aspects of natural history, we incidentally recorded the behaviour of the colony when deprived of food for a week. On the seventh day, as we were about to feed the colonies, we observed the behaviour of "**matriphagy**" in three out of seven colonies. The mothers emerged from the reproductive chambers, raised their pedipalps, and remained passively in this position until their nymphs surrounded them and, suddenly, attacked. That's right: the mothers surrendered themselves as food to their offspring! The nymphs attacked the joints of the mothers' legs, injecting their venom and waiting for the enzymes to act so they could ingest the maternal fluids. The mothers remained completely motionless while being consumed (40 ± 5 min). Immediately afterwards, the mothers' exoskeletons were discarded by the nymphs outside the tree bark that housed the nest. Without the mothers, the nymphs began to hunt cooperatively in the following days (Fig. 7.6).

In addition to our observations, experimental tests were necessary to explain not only how the behaviour occurs, but also its adaptive advantages—the reasons behind this behaviour. Thus, we developed another experimental manipulation, simple on paper but difficult to carry out in practice, as is often the case in Behavioural Ecology. Get used to this if you wish to persist in this beautiful field of knowledge. Our hypothesis, then, was that "**matriphagy enhances the survival of nymphs, increasing their chances of survival and future success.**"

To test this hypothesis, we had to establish, over many months, a series of colonies in the field and laboratory, where we tested the survival of nymphs under three conditions of food deprivation by the mothers.

- In the first group, mothers were isolated with their broods.
- In the second group, nymphs were isolated without their mothers.
- In the third and final group, nymphs were without their mothers but within a well-established colony, with other females and their broods, as well as males and virgin females (Fig. 7.7).

The results showed that the nymphs survive and establish a new colony with all its functions after matriphagy. In other words, when the mother altruistically surrenders herself as food to her offspring, she not only ensures their future survival but also the perpetuation of her genes—that is, her daughters and sons will continue her legacy.

When studying Behavioural Ecology, it is not enough to make behavioural observations, describe ethograms, and test hypotheses as previously presented. At times,

Fig. 7.6 Matriphagy in the social pseudoscorpion, *Paratemnoides nidificator*. In the photograph, at the centre and darker in colour, is a female, the mother of the lighter nymphs surrounding her. In this case, the mother is surrendering herself as food to her offspring. Photo: Everton Tizo-Pedroso [7]

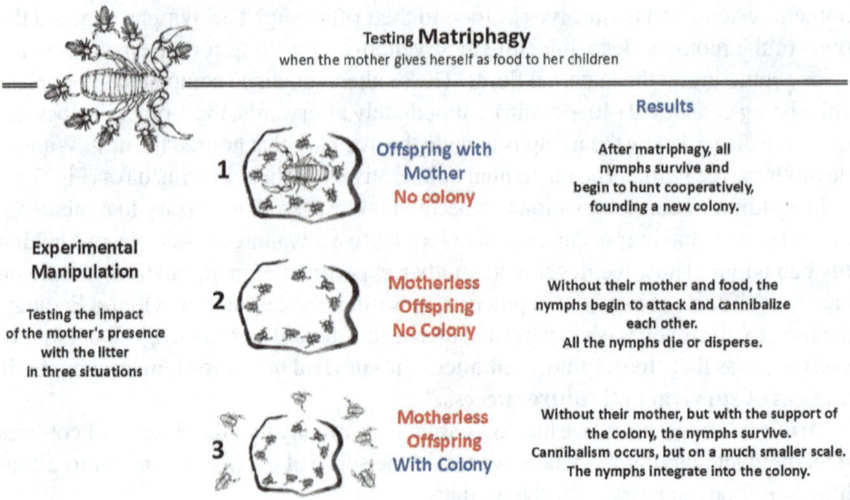

Fig. 7.7 Experimental test of the possible advantage of matriphagy in *Paratemnoides nidificator*. Nymphs, with and without mothers, were exposed to conditions of isolation or colonial support and their survival was monitored over time

phylogenetic comparisons, intra- and intergroup genetic analyses, as well as a range of other studies, have allowed us to jointly outline the behavioural profile of the species. Recently, we published a study involving parasitism and colouration in birds [8]. We considered that haemoprotozoan parasites (e.g., *Plasmodium*) transmitted by vectors can negatively impact host fitness by compromising, for example, important life cycle trade-offs such as sexual selection. In birds, there is a relevant question in this regard: how to balance investment in immune defence with investment in plumage colouration in response to infection—a definitive cost-benefit relationship for the reproductive success of the species. Therefore, understanding why vulnerability to infection varies within a host clade is a crucial task for conservation biology and for understanding macroecological patterns of the life cycle. Here, we studied the relationship between avian life-history traits and climate in the prevalence of *Plasmodium* and *Parahaemoproteus* parasites. We sampled 3569 individual birds belonging to 53 species of the family Thraupidae. Individuals were captured from 2007 to 2018 at 92 sites. We constructed two phylogenetic generalised least squares models with the prevalence of *Plasmodium* and *Parahaemoproteus* as our response variables and the following predictor variables: climate, body size, participation in mixed-species flocks, incubation period, migration, nest height, foraging height, forest cover, and diet. Note the large number of behavioural or life-history variables involved. We found that the prevalence of *Parahaemoproteus* and *Plasmodium* was higher in species from open habitats. Tanager species (Fig. 7.8) with longer incubation periods had a higher prevalence of *Parahaemoproteus*. We hypothesise that these longer incubation periods coincide with the peak abundance of vectors, resulting in a higher probability of

Fig. 7.8 *Tangara cayana* (burnished-buff tanager), one of the study models

infection among adult hosts during their incubation period and among the nestlings. Finally, we found that the prevalence of *Plasmodium* was higher in non-migratory species, in those participating in mixed-species flocks, and in species with an omnivorous or animal-based diet. Note that the study of animal behaviour today goes far beyond simple field observation.

Many of my fellow ecologists underestimate the value of natural history and basic biological observations. It seems they do not understand that, in order to work with ecological interaction networks, for example, I must necessarily know and verify the role of each agent in that network. A network in which I merely record the presence of a beetle on a plant and count it as a true interaction, without testing anything about that relationship, is a flawed network. It is a neat table, but without data of real importance or biological meaning. The beetle may be on the plant because it is a herbivore, because it is a pollinator, or simply because it landed there to rest. And in the latter case, what is the interaction? Yet it appears in "so-and-so's table," just making up the numbers, with no real significance. Now imagine what they think of our ethograms, the description of a species' behavioural repertoire. I agree that, in isolation and without a sequence, it also has little meaning. But in a long-term study, composed of a series of smaller studies, the sum and complexity of what is ultimately obtained is highly significant.

The sum of our efforts in Behavioural Ecology is what allows us to unravel the mysteries of evolutionary life and the adaptive significance of each behaviour. A single study has little real impact compared to the development of a research line, substantiated by a series of publications on a given topic or study model.

Returning to our small arachnid friends, the pseudoscorpions, it was through a series of studies that we were able to demonstrate, for the first time in arachnids, true division of labour involving males as a caste, in a complex study involving ethograms: "*Is there division of labour in cooperative pseudoscorpions? An analysis of the behavioural repertoire of a tropical species*" [9]. In this study, we showed that division of labour is a strategy that maximises foraging and reproductive success in eusocial insects. Although some arachnids exhibit colony structure and social organisation similar to that of hymenopterans, temporal polyethism (division of labour by age within the colony) has only been demonstrated in a few species. The social organisation of the cooperative pseudoscorpion *Paratemnoides nidificator* is similar to that of social spiders, but involves a clear division of labour. Task allocation was investigated experimentally in colonies composed of only one developmental stage (juveniles or adults) or one sex (males or females) through laboratory manipulation. During 44 h of observation, more than 14,000 behavioural repetitions were quantified, distributed across 95 different behaviours and grouped into ten behavioural categories. The results showed that reproductive colonies of *P. nidificator* are maintained by activities based on sex and age. Non-reproductive males and females perform external cleaning of the colony and prey capture. Reproductive females care for the juveniles and build reproductive silk chambers. Nymphs construct most of the moulting chambers and carry out internal cleaning. In the absence of nymphs, male colonies survived for one to two months, while female colonies survived for three to four months. In nymph colonies, work is readjusted so that all maintenance tasks are performed. This was **the first study to clearly demonstrate division of labour in arachnids**. This suggests that specialisation is an adaptive and evolutionarily ancient trait in this social arachnid species. Unlike cooperative spiders, *P. nidificator* possesses physiological (e.g., reproduction, ecdysis, lifespan) and behavioural (e.g., behavioural synchrony or self-organisation) characteristics that allow for task specialisation. "Our little creature" is an *Apis mellifera* without a queen, but with a working male!

Chapter Challenge
Imagine an animal you would like to study, a very specific and unusual behaviour that could be tested both in the wild and in the laboratory. Want an example? Imagine our stingless bees, a *Melipona* species that shows different sizes among its workers. You observe that many of them are more defensive of the nest, while others focus more on parental care. How would you test whether these behaviours have a genetic basis (animal personality?) or are related to age polyethism? A good puzzle for you!

Challenge Objective
To test your ability to seek new knowledge and challenges. To test your skill in problem-solving and hypothesis formulation.

Go Further
See here some short videos to inspire your ideas related to the examples in this chapter.

The first video is about territoriality in dragonflies
https://youtu.be/_FVaa7oxxb8?si=OTA_4ROyP9-9jpgT

The second is about pseudoscorpions
https://youtu.be/mXFH3V_jQxI?si=UZSE-yJho2d5DQTu

The third illustrates the life history of an ant species
https://youtu.be/ALlLxeH2OCI?si=9IU9JE9YgvX6M2xD

References

1. Wheeler WM (1891) The relations of ants to plant life. Bull Torrey Bot Club 18(2):69–82
2. Moura RF et al (2021) Biotic defences against herbivory. In: Del-Claro K, Torezan-Silingardi HM (eds) Plant-animal interactions. Springer, Cham, https://doi.org/10.1007/978-3-030-66877-8_5
3. Byk J, Del-Claro K (2011) Ant–plant interaction in the Neotropical savanna: direct beneficial effects of extrafloral nectar on ant colony fitness. Popul Ecol 53:327–332. https://doi.org/10.1007/s10144-010-0240-7
4. Calixt ES, Lange D, Del-Claro K (2021) Net benefits of a mutualism: influence of the quality of extrafloral nectar on the colony fitness of a mutualistic ant. Biotropica 53:846–856. https://doi.org/10.1111/btp.12925
5. Del-Claro K (1991) Notes on mimicry between two tropical beetles in south-eastern Brazil. J Trop Ecol 7(3):407–410. https://doi.org/10.1017/S0266467400005733
6. Del-Claro K, Tizo-Pedroso (2009) Ecological and evolutionary pathways of social behaviour in Pseudoscorpions (Arachnida: Pseudoscorpiones). Acta Ethol 12:13–22
7. Tizo-Pedroso E, Del-Claro K (2005) Matriphagy in the neotropical pseudoscorpion Paratemnoides nidificator (Balzan, 1888) (Atemnidae). J Arachnol 33:873–877. https://doi.org/10.1636/S03-61.1
8. Penha VAS et al (2022) Host life-history traits predict haemosporidian parasite prevalence in tanagers (Aves: Thraupidae). Parasitology 150(1):32–41. https://doi.org/10.1017/S0031182022001469
9. Tizo-Pedroso E, Del-Claro K (2011) Is there division of labour in cooperative pseudoscorpions? An analysis of the behavioural repertoire of a tropical species. Ethology 117:498–507. https://doi.org/10.1111/j.1439-0310.2011.01906.x

Chapter 8
Applying Animal Behaviour to the Study of Biotic Interactions

Animal Behaviour is now an important complementary tool, or of great utility, for elucidating questions in related scientific fields. In biotic interactions, whether antagonistic—such as predation, parasitism, competition—or mutualistic, as in pollination, seed dispersal, and biotic defences, understanding the behaviour of the animals involved is fundamental. Prey species, for example, evolve strategies to evade or resist predation. Conversely, predators evolve more efficient mechanisms to locate and subdue their prey. The same occurs between parasites and their hosts, between herbivores and plants—a constant genetic arms race that refers to the "Red Queen Hypothesis" [1]. In behavioural ecology, this concept refers to the ongoing evolutionary competition between interacting species within the same environment. The name was inspired by the Red Queen character from Lewis Carroll's *Through the Looking-Glass*, who told Alice in Wonderland that "*it takes all the running you can do, to keep in the same place*."

In behavioural ecology, the Red Queen Hypothesis posits that organisms are engaged in a constant evolutionary race to adapt and survive in an ever-changing environment. For a species to persist, it must continue to evolve and adapt to keep pace with the evolutionary progress of other species with which it interacts, whether positively or negatively. In summary, according to the Red Queen Hypothesis, in a dynamic ecosystem, species are engaged in a perpetual evolutionary race, needing to develop increasingly sophisticated behavioural strategies to avoid predation and parasitism, to forage for resources, to compete for mates, territories, or to engage in any other interactions vital for their survival and reproduction. This concept highlights the importance of interaction and the selective pressure exerted by species upon one another. As one species evolves, it creates selective pressures on the species with which it interacts, which in turn stimulates evolutionary change in those species. This ongoing cycle of coevolution characterises a true "evolutionary arms race" in pursuit of maximum benefit at minimum cost.

In this context, it is important to understand the dynamic role of Animal Behaviour, its techniques, and developments in behavioural ecology that contribute to our

K. Del Claro, *Behavioral Ecology*, https://doi.org/10.1007/978-3-032-13988-7_8

improved understanding of biotic interactions. Biotic interactions, especially those of a mutualistic nature, are perhaps the most important sources of biodiversity on Earth [2]. Understanding biodiversity in natural ecosystems and how it is affected by the Anthropocene (the age of humans, with their drastic impacts on Earth over the past 250 years), by anthropogenic effects on the environment, and by climate change, is a priority for the survival of our own species on the planet. Interaction Ecology today relies on Behavioural Ecology and the tools of Animal Behaviour as pillars that contribute to the elucidation of these important ecological and evolutionary questions.

Let us illustrate this with an example.

8.1 Pollination: Behavioural Ecology Revealing the Nature of a Biotic Interaction

Occasionally, we encounter rare and unexpected events in nature. Often, these events represent unique research opportunities, discoveries that can make significant contributions to science.

As you may imagine, ants are extremely abundant organisms on the Earth's surface, with enormous diversity reflected in a wide range of lifestyles and behaviours. The vast majority are carnivorous, nectarivorous, generalists, and relatively few cut leaves to cultivate the fungi they feed on. They are important predators of other invertebrates, such as butterfly caterpillars, beetles, termites, other ants, and spiders, for example. They can act as seed dispersers or granivores. They may feed on carcasses and contribute directly to the natural recycling of organic compounds. In short, ants do almost everything, which is why you have never, or only rarely, heard of pollinating ants.

Ants have several limitations as pollinators [3], such as their small size, which makes contact with floral structures difficult; their self-grooming behaviour, which can remove pollen grains; limited dispersal ability, among other factors. However, in some cases, this may be different (Fig. 8.1).

During our fieldwork in the Cerrado, after a major fire, we observed some small white flowers close to the ground covered with ants of the genus *Camponotus* walking over them (Fig. 8.2). And no other visitors to the flowers. We then considered that this might be a rare case of ant pollination.

However, between imagining, having an idea, and confirming this hypothesis, there is a long path to be followed in Ecology of Interactions. We found that the tools of Animal Behaviour can play a fundamental role in this process. Animal Behaviour goes far beyond mere observation; it involves experimentation, comparison, and is multidisciplinary.

From a field hypothesis to the reality of a *paper* published in a major journal, the question arises: how did we prove that ants were pollinators of this everlasting from the Cerrado? How did we test the hypothesis that "**ants are pollinators of *Paepalanthus lundii* in the Cerrado**?"

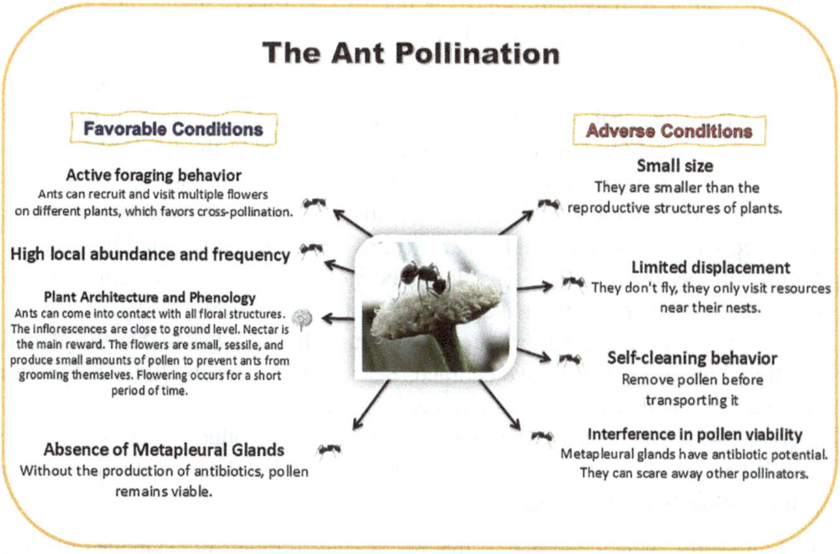

Fig. 8.1 Ants, due to their small size, active self-grooming behaviour, limited dispersal capacity, and a gland that can inhibit pollen germination, are generally unsuitable as pollinators. Their aggressiveness may also deter pollinators from performing their function on plants where ants are present. However, under certain conditions, they can act as pollinators, such as when they are abundant, with flowers that are also plentiful, low-growing, scentless, with little nectar, or unattractive to other insects Adapted from Del-Claro and Torezan-Silingardi [3]

Fig. 8.2 Ant *Camponotus crassus* (Formicinae) on a flower of *Paepalanthus lundii* (Euriocaulaceae), an everlasting

Following what we have outlined so far, there are a series of initial steps that must be taken, which I will enumerate for clarity:

1. Preliminary observation I—a preliminary field analysis should reveal that ants are indeed climbing the plants, visiting the flowers, and contacting the reproductive structures. Or was it all an illusion from a day in the field? What is the behaviour of ants on the flowers?
2. Preliminary observation II—the flowers must be open, with viable free pollen and receptive floral stigmas at the same times when they are visited by ants.
3. Preliminary observation III—ants should be the most abundant and common visitors to these flowers. What is the behaviour of other visitors?
4. Literature—an immediate literature review can help us infer whether the plant in question does or does not have characteristics of (myrmecophilous) pollination by ants, and whether this case is already known.
5. Literature—the literature can tell us whether the reproductive system of this plant is known and what it is, or, if it is not known, what the reproductive system is like in closely related species.

Once this initial stage is completed, the second stage is much more labour-intensive. We need to design experimental field tests, with rigorous controls, that can answer whether ants are indeed effective pollinators of this plant. With this in mind, we considered the following groups, described in Fig. 8.3:

Control: a natural group, *with no interference*. In this group, we marked inflorescences to monitor their fruit production, that is, to determine how many flowers in each capitulum would set fruit/seeds, with free visitation by ants and any other animals.

I. **Treatment**: plants *with no visits from any pollinators*. Will ants or winged insects such as flies, bees, wasps, butterflies set fruit? We needed to measure the spontaneous fruit set in the plants. Pollination tests in this case are difficult to perform, as the flowers are tiny, so an experiment completely isolating the plant could provide a solution.

II. **Treatment**: do plants *visited only by ants* set fruit? How many?

III. **Treatment**: do plants *visited only by animals other than ants* set fruit? How many?

In addition to these observations, we tested pollen viability, stigma receptivity, and the presence of pollen grains on the bodies of ants. Note that we combined animal behaviour with experimental manipulation and tests of plant reproductive biology. Furthermore, anatomical and morphofunctional data of both ants and plants were important.

Analyzing the experiment described to test ant pollination (Fig. 8.3), you should note that, for the **hypothesis to be supported**, the group with the highest fruit set, significantly differing from the other treatment groups, should be **group II—treatment**, with visitation exclusively by ants. This group may not differ only from the control group, where ants have free access to the plants. Furthermore, ants must carry pollen

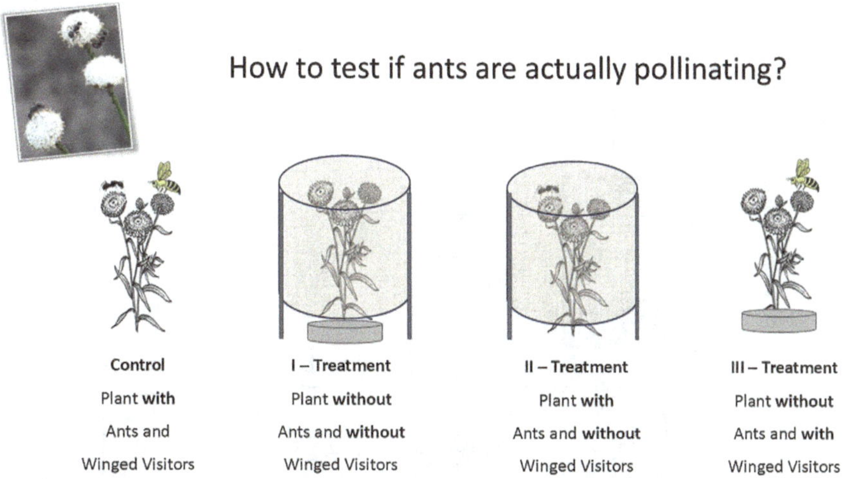

Fig. 8.3 Experimental manipulation test with pollinator behaviour as the background. The **control** group represents the natural state, with pollinators free to visit the flowers. In group **I—treatment**, a plant is covered by a *voile* tent, preventing winged visitors from reaching the flowers, and with a dish containing non-toxic resin at the base, preventing ants from accessing the plant. This represents a plant with spontaneous fruit set, that is, without biotic pollination. In group **II—treatment**, plants are covered by a *voile* tent, preventing winged visitors from reaching the flowers, but allowing free access to ants. In group **III—treatment**, there is a dish with resin non-toxic at the base, preventing ants from reaching the flowers, but allowing free access to other floral visitors

on their bodies (this can be observed under a microscope or stereomicroscope), this pollen must be viable (there are tests for this, such as the acetic carmine test), and they must visit the plants during periods of pollen availability and stigma receptivity (simple tests with hydrogen peroxide can demonstrate this). So, what was the outcome?

The results of this study showed that plants visited by any pollinators (control group) did not differ statistically from plants visited only by ants (group II—treatment). Both groups produced significantly more fruits than the other groups in which ants were excluded (III—treatment) and the group without the action of biotic agents (I—treatment, Fig. 8.4).

This study provided evidence that previous generalisations which overlooked the importance of ants as pollinators were incorrect. It also revealed that the Brazilian Cerrado can offer significant insights into the ant pollination syndrome, as this environment presents unique characteristics related to this association. Thus, this study is highly significant for understanding the ant pollination syndrome and the complex ecological networks present in these arid and dry systems. Moreover, the methodology employed demonstrates a clear and essential use of basic tools from animal behaviour studies for the advancement of Ecology of Interactions.

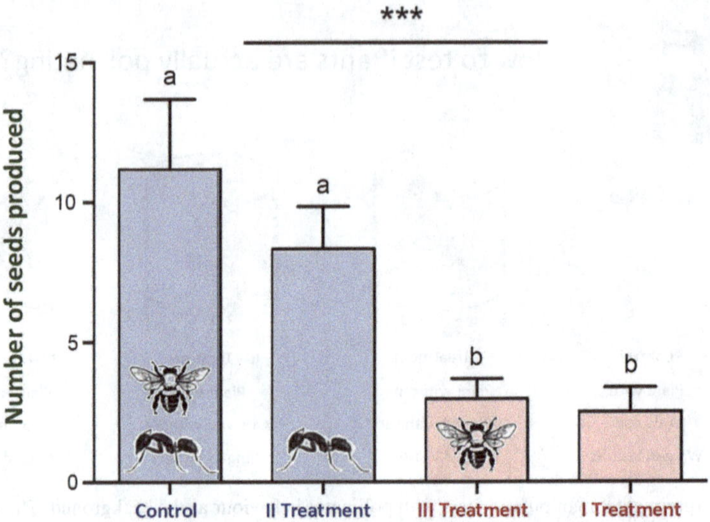

Fig. 8.4 Results of the pollination manipulations carried out in *Paepalanthus lundii*. Plants visited by any pollinators (control group) did not differ statistically from plants visited only by ants (group II—treatment); both produced significantly more fruits than the other groups where ants were excluded and the groups without the action of biotic agents. Identical letters indicate no statistical difference between groups. The opposite is true for different letters, with ***$p < 0.0001$Adapted from [7]

8.2 Facilitation, Ecosystem Engineers, and Animal Behaviour

Interactions between plants and animals play a crucial role in community structure. These mutual relationships can range from positive to negative or neutral for both partners, encompassing everything from predation to mutualism. One type of interaction known as *facilitation* occurs when *the presence of one species modifies the environment in a way that promotes the growth, survival, or reproduction of neighbouring species*. In terrestrial ecosystems, many species that act as facilitators are considered ecosystem engineers. These organisms have the ability to regulate resource availability for other species by causing physical changes to living and non-living materials. Such modifications to the environment increase trophic interactions and species diversity. In addition, ecosystem engineers provide protection against natural enemies and adverse environmental conditions. For example, the construction of shelters such as mines, galls, and leaf rolls increases the survival and growth of insects, affecting the entire associated community.

Choosing a nesting site is fundamental for any species. In the case of ants, many species inhabit fragile, ephemeral, and pre-existing cavities, such as hollow branches. Many ant genera are specialised in nesting in cavities created by wood-boring insects, establishing new colonies or expanding their existing colonies (satellite nests). In the same environment inhabited by these ants, beetles from the families Cerambycidae, Buprestidae, and Curculionidae, for example, excavate galleries within and on the surface of dead wood in vegetation. When abandoned, these galleries and tunnels provide nesting sites for many ant species through intense intra- and interspecific competition.

All right! Ants can use hollows produced by caterpillars of other insects in branches and trunks of plants for nesting. But how do they find these branches?

On one occasion, during our studies in the Cerrado savanna, we observed ants of the genus *Cephalotes* on Pequi branches (*Caryocar brasiliensis*, Caryocaraceae) and near excrement, a mixture of faeces and sawdust resulting from the excavation of trunks by insects. From the literature, we knew that a wood-boring caterpillar (Cossidae: Lepidoptera) had been described as an important pest of Pequi branches and trunks. We also knew that these hollows in Cerrado plants were more frequently produced by beetle larvae. The sawdust with faeces produced by the larvae of butterflies, moths, or beetles accumulates on the trunks, but also falls to the ground. This led us to ask:

Could this mixture of sawdust and faeces be the cue that Cephalotes ants (and possibly other species!) use to locate the hollow trunks where they establish their nests in the Cerrado (Fig. 8.5)?

I must admit that I had never considered that the simple observation of a small pile of faeces mixed with sawdust in the field could result in a paper published in the *Journal of Ethology* [4]. Nature is astonishing, and you must be prepared, when in the field, to know how to observe, to recognise an interesting case for study, and to imagine how to test your hypotheses. This is achieved through knowledge, extensive reading, study, and discussion with your colleagues. Many people go into the field and look, but do not truly see, do you understand? And another thing: never underestimate your curiosity. Do not ignore it; if you see something interesting, seek to study

Fig. 8.5 Nest of *Cephalotes pusillus* in a trunk of *Caryocar brasiliensis* in the Cerrado. The yellow arrow indicates the nest entrance

it, read about it. A great number of excellent studies in Animal Behaviour originate from curiosity, observation, and study. Without studying, without reading the articles being published in your field, without knowing the classic books, you will make very little progress. There is no point in being an outstanding field worker, collecting a thousand and one data points if you do not know what to do with them, if you do not know how to analyse these data and, above all, if you do not know how to interpret the significance of the data you have in your hands.

In this study, after extensive reading, we suggested that the ants could be attracted by chemical/olfactory cues present in the excrement. To test this, several hypotheses were examined; the main one being that "*beetle excrement provides more effective chemical cues for ants than lepidopteran larval excrement,*" since wood-boring beetles are more common and produce more cavities in plants. Through a series of laboratory experiments using ant colonies, sawdust, and sawdust with faeces collected directly from the field, our results revealed that workers of *Cephalotes pusillus* can use beetle excrement as a cue to locate nests in Pequi branches. For the first time, we demonstrated that the selection of a nesting cavity by arboreal ants may be non-random, and is likely guided by a chemical/olfactory cue originating from a facilitating agent, a beetle.

Consider this: although it may seem a very simple question to some, how does a study like this become of interest to an internationally indexed journal? The answer lies in how you develop the study, in which theories you incorporate into your hypothesis testing, and, above all, in how you discuss the "*cost-benefit*" relationship, its direct and indirect impacts, and its conditional results in your manuscript. In other words, it is not enough to find something interesting, nor are field or laboratory tests alone sufficient, nor is it enough to know how to use the tools of Animal Behaviour, I insist… You must study, you must know how to discuss and relate facts and results. For example, in this study involving facilitators (beetles), plants, and ants, we constructed a line of argument that allowed us to state in our final conclusions that studies such as ours increase our "*capacity to understand the ecological and evolutionary mechanisms underlying the behaviour of social insects.*"

8.3 Animal Behaviour in a Study of Territorial Competition

Males of various species possess remarkable visual traits that are selected through interactions between individuals of the same and opposite sex, all in competition for mates. For example, the colourful wings of peacocks and dragonflies, pigmented faces of wasps, and colourful stripes on fish are features that play a key role in male reproductive success. These colour patterns act as signals indicating an individual's quality and are used to communicate with both other males and females. They are reliable signals that help resolve conflicts without the need for physical confrontation. These honest signals reduce costs and increase benefits for all parties involved.

In animals, for instance, in insects, skin colour can reveal much about male quality. Substances such as melanin, carotenoids, and pteridines contribute to this distinctive coloration. Males with more vivid colours generally have a greater ability to protect and defend their territory, to impress females with spectacular displays, and also possess a stronger immune system, which helps them live longer. These attractive colours are sexual ornaments, but producing and maintaining them is energetically costly. Therefore, when we see a male with striking colours, we can trust that he is truly of high quality, as he has invested in his appearance without compromising his health. It is as if the colours honestly reveal his physical condition and territorial status.

Several studies published in the last 10 years indicate that male coloration predicts the acquisition and maintenance of territory in insects such as dragonflies, establishing the basis of their territorial behaviour. However, wing or body pigmentation comes at a cost, such as the need to acquire resources to produce pigments and a higher risk of predation, since more conspicuous animals are more easily detected by predators, and dragonflies have visually acute, agile, and efficient enemies (Fig. 8.6).

In dragonflies of the family Coenagrionidae, males often display sexually dimorphic and conspicuous body colouration, meaning they are much more colourful than females (Fig. 8.7). However, there is little evidence regarding the role of these colour patterns in mediating territorial behaviour.

Aggressive encounters between male dragonflies generally consist of face-to-face agonistic displays, which may involve the pigmentation of the head, eyes, and surrounding areas as a visual signal. Many dragonflies have striking spots in the post-ocular region of the head, which may serve as intrasexual signals of male quality. The dragonfly family with such features, Coenagrionidae, is very common and diverse in Brazilian wetlands. There are several species with highly colourful males, including in the facial region (Fig. 8.7). Thus, after an extensive literature review, our hypothesis was that "**facial pigmentation in dragonflies is a visual communication trait that may indicate territoriality and male superiority.**"

Fig. 8.6 A suiriri, *Tyrannus albogularis*, captured one of several dragonflies, notably *Erythrodiplax latimaculata* in a Cerrado wetland

Fig. 8.7 Male (blue and red) of *Neoneura sylvatica* (Coenagrionidae) in copula with the female (greyish) in a Cerrado wetland

To test this hypothesis, we conducted fieldwork and identified a model species [5]. This species is highly colourful, with pronounced post-ocular colouration in males, easily observed, and abundant for at least one or two months of the year. We sought characteristics that would allow us to test, observe, and re-examine the field study over a suitable time interval, should any errors or unforeseen events occur during the research. We therefore used males of *Tigriagrion aurantinigrum*, which display striking head colouration, with black and orange tones, as our model (Fig. 8.8). We also hypothesised that this colouration should be a selective result of male–male competition. To better understand this trait, we used a technique called static allometry, which compares the size of different body parts. This comparison helped us to understand the relationship between size, head colouration, and its possible evolution due to male competition. In summary, according to sexual selection theory, males develop conspicuous traits to compete for mates. These traits generally increase disproportionately relative to body size, as they are indicators of high genetic quality and signal the ability to win contests. However, this is not always true for all species. The evolution of hyperallometric traits only occurs when the benefits of possessing a trait are greater for larger males. For example, the size of antlers in deer and elk, as well as the horns of Scarabaeidae beetles, or the antennae in Cerambycidae beetles.

In our study, we also investigated whether the facial spots of males influence their territorial behaviour and the outcome of contests. We expected that males with larger (hyperallometric) orange spots would be more successful in contests and spend more time patrolling and fighting to defend their territory. Additionally, we analysed the effect of body size and mass on these outcomes, as they are related to individual physical condition and territorial behaviour.

The results of our study [5] (Fig. 8.8) showed that the orange pigmentation in the post-ocular area of the head of *Tigriagrion aurantinigrum* males allows us to predict

Fig. 8.8 The more orange (orange-golden, no. 1) the face of *Tigriagrion aurantinigrum* males, the greater the chance that the male will stand out and win territorial contests against other males in which dark spots predominate on the face and post-ocular region

which male will win contests for territories and females. In this tiger dragonfly, orange facial colouration proved to be a hyperallometric pattern, indicating that male head colouration is subject to selective forces through male-male competition. Male body size and mass are also good predictors of competitive ability in odonates. However, these traits did not determine the outcome of fights in *T. aurantinigrum*. Here, males with more orange/golden pigmentation on the face were those that won face-to-face agonistic contests ("confrontational displays"), dominating territories and access to females. To our knowledge, this is the first evidence of the use of facial visual signals in resolving disputes between male Odonata.

The importance of a study such as this is that it provides evidence for the significant role that visual traits play in animal communication and conflict resolution. Much attention has been given to the different modes of insect communication, with considerable emphasis on chemical communication. However, here we highlight the importance of studying variation in visual communication as well, and across different taxa.

In this chapter, you have seen three completely distinct studies, all involving ecological interactions—be it mutualism, facilitation, or competition—in which the use of animal behaviour techniques, experimental manipulation, and hypothesis testing were fundamental for satisfactorily addressing the research question. You can see how simple techniques of animal behavioural observation complement laboratory chemical analyses, combined with complex statistical analyses, enabling clear assessment and interpretation of the results. Most importantly, you have acquired all the tools necessary to understand why continuous study and staying up to date are imperative for success in this elegant and appealing field of research. Moreover, you have realised that this field of study is within your reach; nothing is unattainable—on the contrary, it is entirely possible, provided you do what is necessary to achieve this ideal.

Chapter Challenge

In animal–plant interactions, there is a theory known as the "Optimal Defence Theory," which predicts that plants should allocate resources and defences to the parts that are most costly to them, to minimize costs and maximize benefits against herbivore attack. Imagine a plant that possesses EFNs (extrafloral nectaries) to attract predatory ants, which serve as its first line of defence against herbivores. This plant has EFNs on its leaves and at the base of its inflorescences. Imagine an experiment to test whether these plants can induce or manipulate the behaviour of protective ants so that they concentrate more on one part of the plant or another, depending on the intensity of herbivore damage. See Calixto et al. [6] to assist in addressing this question.

Challenge Objective

Now that I assume you have completed all previous challenges, I propose that you go further and work with behavioural ecology, combining behavioural tools with those of ecology. Designing experimental manipulations to test how animal behaviours affect the lives of other organisms, such as plants, and either expand or restrict the biodiversity associated with them, is a highly visible field of science today. This area stands out because we are living in a time of significant loss of biomes and their biodiversity, with climate warming affecting biotic interactions. We need more people engaged in understanding and studying biotic relationships. The aim here is to assess whether you are now able to embark on behavioural ecology.

Go Further

Here I present some videos to help you understand the topics covered in this chapter and their examples.

Pollination by ants
https://youtu.be/DOnLX-GvLSQ?si=gwGzqBUTmReWg5xa
https://youtu.be/30L2jERnckc?si=q_fppi9D7kXgkVA4

Ant–Plant Interactions
https://youtu.be/UDyIm8SWq2s?si=DdjnPks2O0DsUprS
https://youtu.be/WtUf-6pz4wk?si=e5hCCEEoVlmXrrXx

References

1. Van Valen L (1973) A new evolutionary law. Evolutionary theory.
2. Del-Claro K, Torezan-Silingardi HM (2021) Plant-animal interactions. Springer International Publishing, Cham
3. Del-Claro K, Helena Maura Torezan-Silingardi (2020) In search of unusual interactions. A commentary on: 'Pollen adaptation to ant pollination: a case study from the Proteaceae'. Ann Botany 126(3):iv–v. https://doi.org/10.1093/aob/mcaa126

4. Del-Claro K, Anjos DV, Torezan-Silingardi HM (2022) Ant nesting site selection mediated by insects frass. J Ethol 40:31–36. https://doi.org/10.1007/s10164-021-00724-z
5. Vilela DS et al (2017) Colours of war: visual signals may influence the outcome of territorial contests in the tiger damselfly, *Tigriagrion aurantinigrum*. Biol J Linnean Soc 121(4):786–795. https://doi.org/10.1093/biolinnean/blx024
6. Calixto ES, Del-Claro K et al (2023) Variation of the time course of inducibility of indirect defences according to plant tissue and severity of damage. Bull Ecol Soc Am 104(3):e02064. https://doi.org/10.1002/bes2.2064
7. Del-Claro K, Rodriguez-Morales D et al (2019) Ant pollination of *Paepalanthus lundii* (Eriocaulaceae) in Brazilian savanna. Ann Botany 123:1159–1165

Chapter 9
Behavioural Ecology as a Research Field

Human life is full of "*chance events*" that offer us unique opportunities. Every serendipitous encounter and every challenge overcome can lead us down extraordinary paths. These fleeting moments are open doors to new talents and undiscovered passions. Do not underestimate the transformative power of such experiences. Remain open to new opportunities, however small they may seem. Happiness lies in the ability to embrace these "chance events" with gratitude and curiosity. Celebrate fortuitous encounters and allow yourself to be captivated by their magic. Awaken to the wonder of "chance" and embark on a journey of discovery and achievement. Be the protagonist of your own story and turn every "chance event" into an extraordinary moment.

Quite a romanticized text for a book on animal behaviour, isn't it? But there was no other way to tell you that much of what you will discover in your scientific life will arise through opportunities, and if you have sufficient knowledge (read!) to recognise that you are facing a great opportunity, you will be able to seize it.

The study of animal behaviour, through various chance events, changed my life for the better and helped me contribute to the lives, academic success, and personal fulfilment of a significant number of people whom I admire and for whom I have great affection. There were many such chance events that sparked our curiosity, leading us to study and learn more and more, progressing from a *scientific initiation* (very common in undergraduate courses in Brazil, which includes a small scholarship) to a *dissertation* (Master's), *thesis* (PhD), and further publications. It is extraordinary to see so many who happened to cross my path find what they were seeking: an education, a career, or simply the satisfaction of "**I did it too!**"

Since, by chance, we are constantly teaching through examples in this book, I will use a few more to further illustrate how we should study Behavioural Ecology, how this is possible for dedicated students like yourself, and to conclude this introduction to the study of animal behaviour and behavioural ecology.

I am very fond of walking through vegetation—be it savanna, rocky grassland, gallery forest, or rainforest, such as the Atlantic Forest—always alert. I never walk quickly in nature, not least because I live in a place with many snakes, and if you are

K. Del Claro, *Behavioral Ecology*, https://doi.org/10.1007/978-3-032-13988-7_9

Fig. 9.1 Oxyopidae spider feeding on nectar in Malpighiaceae from the Cerrado

not careful, you will get into trouble; pit vipers are found in the field and in university corridors, so be careful.

One day, by chance, I thought I saw a spider drinking nectar from an extrafloral nectary on the leaves of a plant in the Cerrado (Fig. 9.1).

At first glance, I thought it was just a coincidence, as I had always learned that spiders were carnivorous animals, end of story. So, I did exactly what I advise you to do in this book. I returned to the field to observe further, consulted the literature, and spoke with some colleagues. By the end of the week, I was convinced that Oxyopidae, Salticidae, and possibly many other small spiders fed on plant nectar in the Cerrado. In fact, I had already formulated some hypotheses to explain the phenomenon. The hypotheses involved the vulnerability of juveniles. Unable to catch prey—except by cannibalizing a sibling—and needing to obtain food rich in sugars and amino acids easily and predictably, plants become an excellent option. Another aspect of the hypothesis involved the need to obtain water in a dry and hot environment, such as the Cerrado savanna. Nectaries are also a predictable source of water in both time and space. The predictability of a natural resource is something to be richly exploited in nature, as it means survival.

At the time of this discovery, I had a scientific initiation student with me who was very dedicated, humble, intelligent, focused, and collaborative—someone with all the qualities that inspire us to invest in a person's academic future. So, I went into the field with Larissa Nahas, now a professor at the Federal University of Uberlândia, and presented her with the case. From this chance event, the beginning of this excellent professional's career was set in motion.

The first step, as you might imagine, was scientific initiation, followed by a Master's degree. We needed to know which spiders these were; identification is a huge challenge. We had to seek partnerships. Someone who knew spiders, who could help us identify the species, another perspective to broaden our thinking and imagination would be very helpful. And that is what we did. Several months passed. Yes, here everything happens in just a few lines, but you should not imagine that a well-conducted scientific study moves from the discovery of a question to its publication in the blink of an eye. Everything takes time, everything requires effort, everything in science demands dedication. The benefit comes at a cost, as in all things natural or human. Do not be hasty, do not do anything poorly; life's justice catches up with those who do not follow its basic rules.

As soon as we began to know and understand the spiders, a new challenge arose. We needed to describe the feeding behaviour of these spiders and observe and record their interactions with the plant and with other animals. One chance event leads to another, and from these initial studies, we hypothesised that these spiders might go beyond simply feeding on these plants—they might actually benefit them, just as ants do. As efficient predators with behavioural characteristics completely distinct from ants, they could be a force an auxiliary defence for plants against herbivore activity, which would characterise a spider–plant mutualism. Imagine that: spiders protecting plants?

I had never heard of any study proving this. Thus, the literature revealed to us that this would be a fascinating path for someone wishing to pursue a doctorate in a subject involving behavioural ecology.

Yes, it is not only animal behaviour that is involved here. Understanding and describing how spiders feed on nectar constitutes the exclusively behavioural aspect of the study, but, in addition, Behavioural Ecology seeks to uncover the evolutionary causes of this behaviour, that is, how it increases the "fitness," the adaptive value of spiders. Such a peculiar behaviour must enhance the chances of survival and subsequent reproduction of these spiders.

On the other hand, we know that plants, as primary producers, are consumed by herbivores and that, if herbivory is intense, their chances of reproduction may be reduced. Loss of photosynthetic area leads to a reduction in resources available for the production of flowers and fruits. The "Optimal Defence Theory" (ODT) posits that plants should invest in the defence of their most valuable parts, such as new leaf buds, flower buds, flowers, and fruits. Thus, if the relationship with spiders is truly beneficial for both these predators and the plants, the cost of producing nectar should be offset by the benefit of protection provided by spiders against herbivores.

The tools of Animal Behaviour are intertwined here with the story of a possibly coevolutionary mutualism between ants and plants with extrafloral nectaries. And we proposed to include spiders in this relationship.

Behavioural Ecology is multidimensional; it *explores the animal's universe and its relationships with other organisms and the environment, seeking to understand how the outcomes of these interactions affect the adaptive value of the animal and the other organisms involved.*

But did spiders, in fact, ingest the nectar? Imagine if a journal reviewer, the rigorous and demanding type, would not question this? A photograph alone would not suffice; in this case, we had to provide proof. Thus, part of the study involved observing, collecting the spiders, and demonstrating that their stomachs contained plant nectar (fructose) in an irrefutable quantity, so that the reviewer could not claim that the sugar in the spider's stomach originated from an ant it had eaten.

Ah, you had not considered that?

Indeed, when working in science, one must ensure there are no loose ends!

A loose end in the construction of your "scientific knitting" will unravel the fabric, weaken your arguments, and may even undermine your working hypotheses. Therefore, we had to turn to the literature. With a study published in 1972, we learned part of what we needed: the "anthrone method," using cold sulphuric acid (anthrone) to test for the presence of nectar (polysaccharides) in the digestive tract of mosquitoes [1]. Based on this and other articles, we developed a graph of fructose concentration values in standard solutions and their respective absorbance values. In this way, we determined the fructose concentration in all spider samples using a simple equation. In this part of the study, the spiders tested belonged to seven families: Anyphaenidae, Araneidae, Oxyopidae, Pisauridae, Salticidae, Theridiidae, and Thomisidae. A total of 265 (88.04%) individuals among all those tested, including males, females, and juveniles, showed positive results for fructose ingestion. In many families, all individuals captured had fructose in their stomach contents, and in most cases, at levels above 50%, which clearly rules out chance. The results found in this study corroborate other studies suggesting that nectar feeding is a common activity among spiders [2].

Thus, we had a *hypothesis* in Behavioural Ecology:

"Spiders feed at extrafloral nectaries and act as

biotic defenders of plants in a facultative mutualism"

The acceptance of this hypothesis depended on the confirmation of two facts (Fig. 9.2):

1. spiders do indeed feed on nectar from EFNs;
2. plants benefit from the presence of spiders on them, with reduced foliar herbivory and/or increased fruit and seed production.

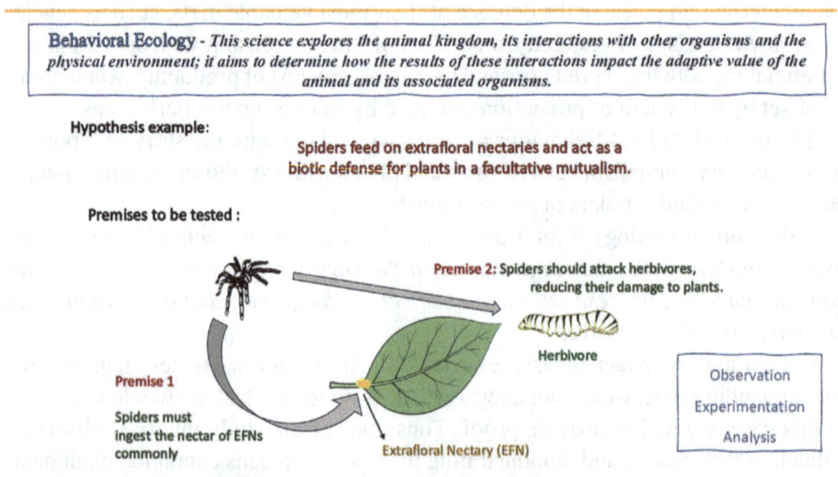

Fig. 9.2 Behavioural Ecology—example of the formulation and testing of a hypothesis through the analysis of experimental stages and the testing of premises

Once it was confirmed that spiders feed at nectaries, and knowing from the literature that extrafloral nectar is rich in water, sugars, amino acids, and even plant-derived lipids, we moved on to the second premise. We needed to test whether spiders, when climbing plants in search of nectar, had any impact on the insects, especially the herbivores, associated with those plants. Although this may seem a straightforward experimental manipulation, one might think of simply having plants with and without spiders and measuring the intensity of leaf area loss over time; however, other factors are involved and have a direct and significant influence on the results.

In Behavioural Ecology, we deal with questions related to the adaptive value of organisms, measuring directly or indirectly their ability to survive longer and, consequently, increase their offspring, thus contributing to the genetic pattern of future populations. But, as organisms are constantly interacting with both the physical and biotic environment, in order to understand their role in structuring natural communities and in the conservation of biodiversity, we must go further. We need to understand the consequences of their decisions, the behaviours of these animals towards other organisms with which they interact and the environment they inevitably modify. In our example, it would not be enough to simply isolate the plants from the presence of spiders; we would also need to know, for instance:

a. how many days it takes for a plant to be rediscovered and recolonised by a spider, an essential behavioural piece of information for us to determine how often we would need to visit the field to remove spiders from treatment plants;
b. if plants with EFNs are visited mainly by spiders and ants, then we would need to know the isolated impact of ants, spiders, both together, and the simultaneous absence of both on the plants.

Thus, all these experimental variations had to be planned and implemented under natural conditions in order to test the second premise that would validate or not our hypothesis [3]. It was necessary to determine whether plants with free access to ants and spiders would experience the same impact from herbivore activity as plants visited only by spiders or only by ants. Furthermore, we needed to compare these plants with those without access to ants and spiders.

Plants with both spiders and ants were necessary to determine the complementary effect of two distinct predators on plant reproductive success and on herbivore activity. Plants with only one type of predator, either spiders or ants, would reveal the isolated effect of each predator group. In turn, plants without predators, with the total exclusion of these animals, would allow us to determine comparatively whether their presence or absence on the plants had any effect on herbivory, plant survival, and reproduction.

Therefore, plant characteristics can also be important in a study of Animal Behaviour or Behavioural Ecology and, in this case, rates of leaf area loss before and after each experimental manipulation had to be recorded and compared. The rate of fruit production from floral buds produced is essential information. In addition, of course, we needed to quantify and qualify the herbivores, spiders, and ants on the plants (Fig. 9.3).

This set of studies enabled us to demonstrate that multiple predators often have effects on populations of their shared prey that cannot be predicted by simply summing the effects of each predator individually. The results showed that the presence of

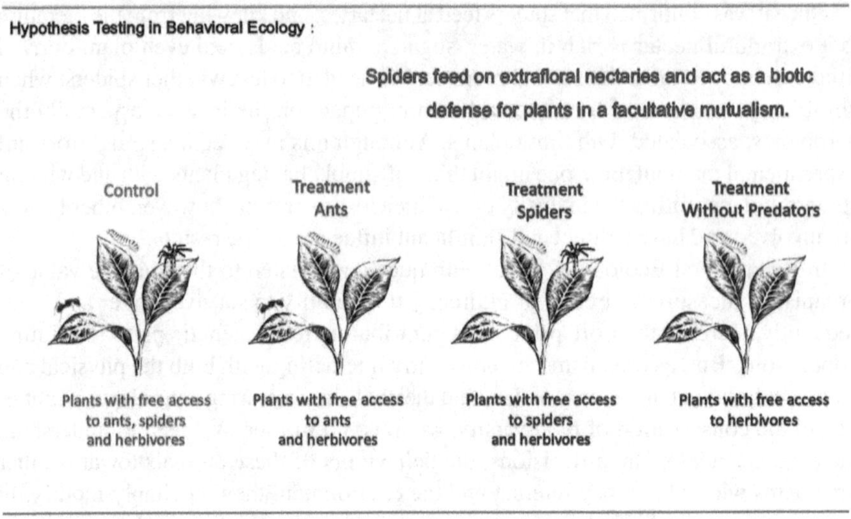

Fig. 9.3 Example of experimental manipulation in Behavioural Ecology. A control group, in its natural state, is here compared with three field treatments, removing one or the other, or all types of predators from the plants. Subsequently, data on foliar herbivory, number of herbivores, fruit and seed production are monitored

ants reduced the abundance and richness of spiders, but did not affect the abundance and richness of ants. Only the removal of ants resulted in a significant increase in the abundance and richness of herbivores. However, herbivory was also reduced by the presence of spiders. We also found a significant interaction effect between ants and spiders on herbivory, indicating an emergent effect of multiple predators. Overall, herbivory was lower. This study highlighted the importance of assessing the effect of the predatory fauna as a whole, rather than focusing solely on a specific group, on herbivory. It demonstrates how significant the contribution of the tools used in Animal Behaviour is to our understanding of ecology and the implications of natural selection on the adaptive value of organisms.

This type of study once again illustrates the importance of in-depth literature review, of seeking explanations and clues in the scientific literature so that we can elucidate relevant questions in the field of science in which we work. Behavioural Ecology is today one of the most multidisciplinary areas of the biological sciences, with significant contributions to a better understanding of how the biodiversity of ecological interactions that sustain life on our planet is structured and maintained.

Chapter Challenge

Draw up a step-by-step flowchart, a basic outline for developing a Behavioural Ecology study.

Challenge Objectives
As you can see, once again I am pushing you towards planning. The aim here is for you to know how to outline before writing, to develop a flowchart of the steps to follow in planning a project and research, but now with the focus entirely on Behavioural Ecology.

Go Further
Watch here a video about our study on the complementary effects of spiders and ants on the plants they visit:

https://youtu.be/WtUf-6pz4wk?si=5rga4OalG8103nzn

Find further examples and additional tools in my previous book, available free of charge at: *The Science We Do.* https://www.cienciaquenosfazemos.org/

References

1. Van Handel E (1972) The detection of nectar in mosquitoes. Mosq News 32:458
2. Nahas L, Gonzaga MO, Del-Claro K (2017) Wandering and web spiders feeding on the nectar from extrafloral nectaries in neotropical savanna. J Zool 301:125–132. https://doi.org/10.1111/jzo.12400
3. Nahas L, Gonzaga MO, Del-Claro K (2012) Emergent impacts of ant and spider interactions: herbivory reduction in a tropical Savanna tree. Biotropica 44:498–505. https://doi.org/10.1111/j.1744-7429.2011.00850.x

Chapter 10
Teaching Practice in Behavioural Ecology, a Conclusion

Teaching through examples, repeating practical exercises while assigning the appropriate value and time to each stage, leads to improvement. The introduction of innovations that come from reading, information, and contact with other researchers refines our ideas and transforms our knowledge, raising its scientific and intellectual standard. Our dedication, attention to detail, and care in our daily scientific work are what earn us a respectable reputation in the academic world. All this effort gives us credibility.

Take it slowly, do not put the cart before the horse; if you do, it will not work out. Everything comes in its own time, with *dedication*, tireless *practice*, and *patience*. Yes, **patience**, because many things will go wrong and will need to be redone. Persistence is essential. That is the nature of the game. So, what better way to conclude our scientific discussion on animal behaviour and behavioural ecology than by proposing a skills test—a practical activity in behavioural ecology that will allow you to put to the test all the new skills you have acquired from our scholarly conversation?

Is everyone ready? Let's begin!

This will be a ***challenge chapter***, as I will ask you to develop a research project based on field observation. However, this exercise will require commitment, perseverance, dedication, and a genuine desire to learn. It is up to you! The decision is yours, as this reading is not an obligation; it never was—it is a choice.

That is how it is in science: we make choices.

Therefore, we must always think carefully about the choices we make, as all of them take time, and time is the one thing we cannot regain.

This project should be comprehensive, about eight pages long, and include:

1. a title that reflects a summary of your hypothesis/proposal;
2. the selection of your scientific name;
3. keywords that do not repeat words in the title;
4. a 150-word abstract;
5. an introduction, moving from the general to the specific, culminating in the hypothesis;

K. Del Claro, *Behavioral Ecology*, https://doi.org/10.1007/978-3-032-13988-7_10

6. a hypothesis based on clear, simple arguments and questions;
7. a detailed methodology, from study site to likely statistical analysis;
8. a timeline covering all stages;
9. references, including at least one from an international journal.

So, this is a rather complex and highly instructive skills exercise.

Let us move on to the problem, a real case. See Fig. 10.1:

In the images in Fig. 10.1, we see a wasp, possibly *Bicyrtes angulatus* (Hymenoptera, Crabronidae, Bembicini), a sand-digging wasp. The wasp arrives, opens a small hole in the sandy soil of the Cerrado savanna, and flies off. It then returns with a prey item, a bug (Hemiptera: Alydidae), held between its legs. The wasp resumes digging. This takes about 10 min. She leaves the hole alone and closes the entrance. She brushes her legs as if sweeping, disguising the entrance to the hole, and flies away. The entire process takes between 30 and 45 min.

However, you are intrigued by the observation and remain at the site. Other individuals of the same wasp species arrive, all within a sandy area about 10 m long and 4 m wide. Around 50 wasps come and go (are they the same individuals?). They open holes, repeating the process that began around 9 am. You use your net to capture 15 of these wasps to examine their prey, always Alydidae, and all adults. You release the wasps and continue observing. Suddenly, at 15:37, all activity ceases. You are standing in the intense sun and heat of June in the savanna. Although it is winter and dry, under the sun the Cerrado savanna is scorching, even in winter.

In your observation area, you also noticed some other small wasps, some Crabronidae, Sphecidae, and the occasional Pompilidae. There were also some small Phorid flies. Ants of various species and sizes were present.

Fig. 10.1 Wasp of the genus *Bicyrtes* sp (Hymenoptera, Crabronidae, Bembicini). Description, in the text, of the behaviour observed in the figure below

Our challenge is for you to use these images and information to develop a "Scientific Initiation" project with the potential of a master's dissertation, employing Animal Behaviour tools. However, we want you to go further, and make this a Behavioural Ecology project, as you will need to test in the field and/or laboratory the adaptive value of a particular behaviour, which will underpin your main hypothesis. You will find very little about these wasps in the literature. A study based on these observations could indeed be novel—a fine challenge for anyone truly interested in working with behaviour and behavioural ecology. Remember, learning the techniques, the methods, and training is just the initial, basic step for those who wish to go further. One cannot study a whale without first knowing the sandflies on the shore.

I hope you have enjoyed this conversation of ours, for although it is a monologue, I believe in a dialogue between my words and your mind. I hope this text has inspired you, helped you to learn, to understand, and to define your interests—or not—in this fascinating and important field of science. Consider me a friend, a partner with whom you can always engage in dialogue on social media whenever possible.

I love working with Behavioural Ecology for the challenges and rewards it offers, for its beauty, for the friends I have made and continue to make, for having taught me to see wonderful and beautiful things both in tiny creatures, to which we often pay little attention, and in large and well-known animals.

But above all, because this field has allowed me to contribute to the growth, success, and happiness of many people.

With warm regards,
Kleber Del Claro

Glossary of Useful Terms in Animal Behaviour

Adaptation any change in the structure or function of an organism (or one of its parts) that makes it better suited to its environment and/or to the functions it performs within it (*Adaptation*).

Agonistic aggressive or combative behaviour (*Agonistic*).

Agonistic behaviour behaviour displayed by animals during competitive interactions, such as threats, aggressive displays, fights, etc.

Alarm signal a signal indicating the presence of danger (*Alarm signal*).

All occurrence sampling also known as "*all occurrence sampling*", this is a recording technique in which animal behaviour is sampled ad libitum (*ad libitum*). All behavioural acts performed by the animal are recorded.

Altruism behaviour in which the action harms the individual itself but benefits others, usually its relatives (*Altruism*).

Animal behaviour any act performed by an animal, whether or not it is perceptible to the human sensory world. Everything an animal does or refrains from doing (*Animal behaviour*).

Animal welfare a branch of ethology that studies how to improve the living conditions of animals in captivity (*Animal welfare*).

Aposematism conspicuous patterns, colours, behaviours or even sounds adopted by harmful organisms to warn predators of their potential danger (*Aposematism*).

Asexual reproduction reproduction occurring without the benefit of sexual union of gametes (fertilisation) (*Asexual reproduction*).

Batesian mimicry mimicry in which a palatable species (mimic) resembles an unpalatable species (model) to deceive predators (*Batesian mimicry*).

Behaviour an action or reaction of an organism in response to internal or external stimuli.

Behavioural catalogue (behavioural repertory) a listing of some of the known behaviours of an animal, also referred to as "*Behavioural Repertory*". See *Ethogram* (*Behavioural repertory*).

K. Del Claro, *Behavioral Ecology*, https://doi.org/10.1007/978-3-032-13988-7

Behavioural category category of an ethogram (*Behavioural category*).

Behavioural ecology the science that studies animal behaviour considering its ecological and evolutionary bases. Through experimentation, it seeks to demonstrate the evolutionary (adaptive) causes of behaviours (*Behavioural Ecology*).

Behavioural repertoire description of all behaviours an animal is capable of performing. See *Ethogram* (*Behavioural repertoire*).

Behaviourism a school of psychology that proposes that the study of behaviour should be conducted only through observable actions (*Behaviorism*).

Benefit an improvement in survival and reproductive success (gain) as a result of a behaviour (*Benefit*).

Charles Robert Darwin father of the theory of natural selection, biologist, naturalist, geologist, and evolutionist who radically changed the way humanity understands science and how humans fit into global ecology.

Chronobiology the scientific study of the "biological clock" of organisms, i.e., how they respond physiologically and psychologically over a regular time interval (*Chronobiology*).

Coevolution the occurrence of genetically determined traits (adaptations) in two or more species, shaped by the interactions between the organisms involved, subject to the action of natural selection (*Coevolution*).

Communication a signal from one organism that alters the behavioural pattern of another (*Communication*).

Communication behaviour the use of signals, vocalisations, or other means to transmit information between individuals of the same species.

Comparative approach (comparative method) a procedure used to test evolutionary hypotheses based on systematic comparisons (predictions) between species with known evolutionary relationships (*Comparative method*).

Comparative psychology a branch of psychology that minimises the evolutionary perspective, focusing primarily on associative learning (*Comparative psychology*).

Competition the contest for the same resource by two or more species. It occurs when the resource is generally insufficient to meet the combined needs of these species (*Competition*).

Conflict of interests a situation in which the individual interests of different individuals come into conflict, resulting in competitive or antagonistic interactions.

Conspicuousness/visibility the ability of an animal to be highly visible or striking (*Conspicuity/Conspicuously*).

Cooperation behaviour in which individuals work together for mutual benefit, generally increasing their reproductive fitness or resource acquisition.

Copulation reproductive behaviour resulting in the deposition of sperm from the male into the female's reproductive tract (*Copulation/Mating*).

Critical period the age during which a particular type of learning must occur, or when it occurs much more readily than at other times (*Critical period*).

Cryptic or camouflaged coloration the resemblance of an animal to some part of its environment. This occurs when an animal's colour pattern blends with the environmental background in which it is found (*Cryptic colouration/Camouflage*).

Direct fitness genes passed by an individual to the next generation through its own reproduction, as manifested in its descendants.
Disruptive coloration coloration that breaks up the outline of an animal's body (*Disruptive coloration*).
Display a stereotyped action used as a communication signal by certain individuals.
Dominance behaviour behaviour displayed by animals within a social hierarchy, demonstrating superiority over other individuals.
Dominance hierarchy an ordered ranking of individuals within a group, based primarily on the outcomes of aggressive encounters (*Dominance hierarchy*).
Ecology the science that studies the relationships of organisms with their biotic and abiotic environment, how organisms affect and are affected by the environment, referred to as the "economy of nature" by Ernest Haeckel (*Ecology*).
English the language you must learn to speak if you wish to become a scientist.
Ethogram a set of descriptions of the basic characteristics of a species' behavioural patterns. See *Behavioural catalogue* (*Ethogram*).
Ethology the science that studies animal behaviour (including human behaviour) (*Ethology*). Branch of biology that studies animal behaviour in their natural environment. Throughout history, several researchers have contributed to this field of study. Here are some of the leading figures in ethology:

1. Konrad Lorenz: regarded as one of the founders of ethology, Lorenz was awarded the Nobel Prize in Physiology or Medicine in 1973. He is known for his work on *imprinting* and for the concept of the critical period in animal development.
2. Nikolaas Tinbergen: another pioneer of ethology, Tinbergen received the Nobel Prize in Physiology or Medicine in 1973, together with Lorenz and Karl von Frisch. He is renowned for his studies on releasing stimuli and fixed action patterns.
3. Karl von Frisch: shared the Nobel Prize in Physiology or Medicine in 1973 with Lorenz and Tinbergen. He is famous for his research on bee communication and language, specifically the waggle dance of bees.
4. Jane Goodall: renowned primatologist and pioneer in the study of chimpanzee behaviour. Her work revolutionised our understanding of primates and their close kinship with humans.
5. Dian Fossey: also a primatologist, known for her work with mountain gorillas in Rwanda. She studied the social life and behaviour of gorillas, and her research helped raise awareness of the importance of conserving these species.
6. Frans de Waal: primatologist and ethologist known for his work with primates, especially bonobos and chimpanzees. He studied social behaviour, empathy, and morality in primates, challenging the view that these traits are uniquely human.
7. Edward Osborne Wilson: known as E. O. Wilson, is an entomologist and sociobiologist who also made significant contributions to ethology. He is famous for his studies on ants and for his theories on sociobiology and biodiversity.

Eusocial/true social term applied to social insects whose individuals exhibit the following characteristics: cooperative brood care, generational overlap, and reproductive division of labour. They generally live in a single nest (*Eusocial*).

Evolution change in heritable traits through the replacement of genotypes within a population. In general, this change results from natural selection, mutation, or genetic drift. Evolution should not be understood as synonymous with progress, but simply as change or modification (*Evolution*).

Evolutionarily stable strategy (ESS) a strategy such that, if all members of a population adopted it, no alternative strategy could invade the system (*Evolutionary stable strategy*).

Fitness (adaptive value) a measure of the genes passed by an individual to the next generation, often quantified by the number of offspring produced that survive to reproductive age.

***Fixed action pattern* (FAP)** genetically programmed behaviour (also called stereotyped behaviour), usually triggered by a stimulus (sign stimulus). More recently renamed as "modal action patterns" (MAP) (*Fixed action pattern*).

Focal animal a technique used to observe a single individual or a group of individuals ("focal group") that, similarly to "*snapshots*", allows for behavioural censuses of individuals at regular time intervals. In a group, the focal individual can be changed every minute, for example (*Focal sampling/Focal animal sampling*).

Food web representation of multiple pathways of energy flow through populations in a community, taking into account that each population shares resources with others (*Food net, food web*).

Foraging behaviour behaviour related to the search for, acquisition, and consumption of food.

Game theory analysis of behavioural decisions in which outcomes depend on the behaviour of other interacting individuals (*Game theory*).

Generalist a species with a broad preference for foods or habitats (*Generalist*).

Habitat the space or location where an animal or plant typically lives, often characterised by a dominant plant form or physical feature.

Habituation the simplest form of learning, in which an animal is exposed to a stimulus without reward or punishment and subsequently responds in some way to the stimulus (*Habituation*).

Herbivore an organism that consumes living plants or parts thereof (*Herbivore*).

Honest signal a characteristic or behaviour that provides accurate information about an individual's quality or fitness, increasing the effectiveness of sexual selection or competition for resources.

Host a living organism on or within which a parasite or commensal resides (*Host*).

Imprinting a rapid form of learning in which an animal acquires a specific response that is retained for life.

Inclusive fitness the sum of an individual's direct and indirect adaptive value.

Indirect fitness genes passed by an individual to the next generation indirectly, primarily by assisting in the rearing of relatives' offspring.

***Individual fitness* /"Fitness"** future contribution to the "*pool*" of the next generation that can be measured by the offspring, by the number of viable descendants left by the individual in the population (*Adaptive value*).

Innate behaviour behaviour developed without the need for learning or prior experience. It is primarily determined by genetics.

Instantaneous sampling also known as "*snapshots*", "*instantaneous sampling*" or "*photographs*", this is a behavioural recording technique in which the behaviours of one or more individuals are observed at regular time intervals. For example, this technique can be applied to observe a colony of social wasps or ants, where the behaviour of all individuals is recorded every five minutes.

Instinct a behavioural pattern developed in most individuals that elicits a functional response the first time an action is performed (*Instinct*).

Intraspecific competition competition between members of the same species (*Intraspecific competition*).

Interspecific competition competition between individuals of two or more species (*Interspecific competition*).

Kin selection differential reproduction among related lineages based on genetic variation and social behaviour (*Kin selection*).

Learned behaviour behaviour acquired through experience, observation, or interaction with the environment.

Learning a lasting and usually adaptive change in an individual's behaviour resulting from life experience (*Learning*).

Macroevolution evolutionary changes that occur over long periods of time and generally involve alterations in many characteristics (*Macroevolution*).

Mathematics in the near future, a new way of writing biology (Biomathematics).

Mate choice selection of a mate based on the characteristics of its phenotype or its territory (*Mate choice*).

Mating behaviour behaviour exhibited by animals to attract mates and reproduce, which may include courtship displays, singing, dancing, among others.

Microevolution small evolutionary changes that occur over short periods of time (*Microevolution*).

Migratory behaviour the periodic movement of an animal from one region to another, generally in response to seasonal or resource changes.

Mimic an organism adapted to resemble another organism (*Mimic*).

Mimicry the resemblance of one organism (animal) to another animal (model). This similarity evolves to deceive predators or prey by confusing the mimic with its model (*Mimicry*).

***Model* (in mimicry)** an organism, usually unpalatable or harmful to predators, that is imitated by the mimic (*Model*).

Model organism the model or animal on which research is focused (Model).

Monogamy a mating system in which each individual mates exclusively with one individual of the opposite sex, usually involving a strong and lasting pair bond (*Monogamy*).

Müllerian mimicry mutual resemblance between two or more unpalatable species to reinforce predator avoidance (*Müllerian mimicry*).

Mutation a change in the structure of a gene or set of chromosomes is typically random. Mutations can lead to inherited variations (*Mutation*).

Mutualism a relationship between two species that benefits both. It may be obligatory or facultative (protocooperation) (*Mutualism*).

Natural selection variation in gene frequency within a population through the survival and differential reproduction of individuals carrying certain alleles, to the detriment of others (*Natural selection*).

Neurobiology the science that studies behaviour with particular emphasis on the mechanisms of the nervous system rather than evolutionary aspects (*Neurobiology*).

Niche the functional role of a species in relation to other species and its physical environment (*Niche*). The specific role of a species or population within an ecosystem, including its interactions with the physical environment and other organisms.

Omnivore an organism with a broad diet, including both plants and animals; specifically, an organism that feeds at more than one trophic level (*Omnivorous*).

Optimal foraging a set of actions, including broadening or narrowing of diet breadth, by which an organism maximises food intake per unit time or minimises the time required to obtain necessary food. It is now recognised that foraging may also apply to other resources such as nest sites, mates, hunting grounds, etc. (*Optimal Foraging*).

Optimum a narrow range of environmental conditions to which an organism is best adapted (*Optimum, Optimal*).

Paleoethology the study of behaviour through fossils (*Paleoethology*).

Parental care behaviour behaviour exhibited by parents to care for and protect their offspring, which may include nest building, feeding the young, protection against predators, etc.

Parental investment the investment by parents in an offspring or group of offspring that reduces their ability to assist other offspring (*Parental investment*).

Photoperiod the duration of a period of light, such as the length of time in a 24-hour cycle during which daylight is present (*Photoperiod*).

Phylogeny the evolutionary history of a specific group of organisms that shows the genetic links between ancestors and descendants (*Phylogeny*).

Polyandry a mating system in which a female mates with more than one male at the same time or in rapid succession (*Polyandry*).

Polygamy a mating system in which a male mates with more than one female, or a female mates with more than one male at the same time or within a short period (*Polygamy*).

Polygyny a mating system in which a male mates with more than one female at the same time or within a short period (*Polygyny*).

Polymorphism the occurrence of more than one distinct form of individuals or genotypes within a population (*Polymorphism*).

Predator an animal (rarely a plant) that kills and eats other animals (*Predator*).

Preliminary observations initial observations regarding a given model organism (*Initial, preliminary observations*).

Prey an organism consumed as a source of energy (*Prey*).

Promiscuity mating with many individuals within a population (*Promiscuity*).

Reciprocity reciprocal altruism in which an individual returns a beneficial action previously received from another individual (*Reciprocity*).

Refuge a place where a species or community can persist in the face of environmental change affecting the rest of its distribution (*Refuge, shelter*).

Releaser a sensory stimulus that triggers a fixed action pattern (*Releaser stimulus/Stimulus release*).

Reproductive effort (investment) allocation of time or resources, or the act of taking risks, in order to increase fecundity (*Reproductive worth, investment*).

Reproductive strategies the set of behaviours and physical characteristics that individuals display to maximise their reproductive fitness, which may include mating strategies, parental care, competition for resources, among others.

Resource something in the environment required by an organism for its maintenance and growth, and which is consumed in the process of its use (*Resource, source*).

Retro-orientation the impression that the head is on the opposite side to where it should be (for example, the tail) (*retro-orientation*).

Searching image a perceptual mechanism used by a visually oriented predator to search for prey. When a predator can distinguish between two or more prey colour patterns and selectively searches for one of them in the environment, it is said to have formed a searching image for that prey (*Searching image*).

Selfish act (behaviour) a behavioural action that benefits the performer but harms the recipient or those affected by its outcome (*Selfish act or behaviour*).

Sequence sampling also known as "*sequence sampling*", this is a technique for recording behavioural acts in which a sequence of events is observed within a defined time interval. This technique is typically used to study a specific behavioural event, such as copulation.

Sex ratio in a population, the ratio of the number of individuals of one sex to those of the other sex (*Sexual ratio, Sex ratio*).

Sexual dimorphism a condition in which males and females of a species differ in appearance (*Sexual dimorphism*).

Sexual reproduction reproduction through the union of two gametes (fertilisation) to form the zygote (*Sexual reproduction*).

Sexual selection selection of individuals of one sex by individuals of the opposite sex based on phenotypic characteristics (usually appearance or behaviour) (*Sexual selection*).

Signal an environmental feature perceived by an animal that can be used as information by other animals.

Sign stimulus/releaser a stimulus that triggers a fixed action pattern (*Sign stimulus*).

Social behaviour any direct interaction between individuals of the same species, generally related and living in a group (*Social behaviour*). It occurs among individuals of the same species, involving social interactions such as communication, cooperation, or competition.

Society a group of individuals belonging to the same species and organised in a cooperative manner (*Society*).

Specialisation adaptation in form or function that enables an individual to perform particularly well in a restricted set of habitats, resources, or environmental conditions (*Specialisation*).

Speciation evolutionary process that splits a population into two, reproductively isolated from each other (*Speciation*).

Stimulus a characteristic of the abiotic or biotic environment that elicits a response in a receptor (*Stimulus*). Any detectable change in the environment that provokes a behavioural response.
Symbiosis any relationship between two organisms of different species (*Symbiosis*).
Taxonomy the science of classifying organisms (*Taxonomy*).
Territorial behaviour behaviour in which an animal defends a specific area as its territory against intruders of the same species.
Territory any area defended by one or more individuals against intrusion by other individuals of the same or different species (*Territory*).
Thanatosis tonic immobility, feigning death (*Thanatosis*).
Variation difference between individuals within the same species (*Variation*).
Vicariance a process in biogeography and evolutionary biology in which the geographical distribution of a species is split or fragmented due to the formation of barriers or changes in the landscape. This separation may result from geological events, such as the formation of mountains, rivers, or other physical barriers, or from environmental changes that lead to the isolation of populations.
Warning coloration coloration that signals danger, toxicity, unpalatability, or aposematism (*Warning coloration/Aposematic coloration*).